THE OPEN UNIVERSITY

Science : A Second Level Course

Structure, Bonding and the Periodic Law

6   Oxidation States and the Typical Metals

7   Classical Bonding Theories

Prepared by an Open University Course Team

…iversity Press

A NOTE ABOUT AUTHORSHIP OF THIS TEXT

This text is one of a series that forms part of the Second Level Science Course, *Structure, Bonding and the Periodic Law*. The other components are a series of television and radio programmes, home experiments and a summer school.

The Course has been produced by a team, which accepts responsibility for the Course as a whole and for each of its components.

THE S25- COURSE TEAM

Chairman and General Editor

L. J. Haynes

Unit Authors

C. J. Harding
D. A. Johnson
Joan Mason
Jane Nelson
R. A. Ross

Editors

Jacqueline Stewart (*Editor*)
Maggie Harris (*Editorial Assistant*)

Other Members

A. Clow (*BBC*)
R. M. Haines (*Staff Tutor*)
R. R. Hill
D. S. Jackson (*BBC*)
G. W. Loveday (*Staff Tutor*)
G. D. Moss (*IET*)
D. R. Roberts
R. C. Russell (*Staff Tutor*)
N. A. Taylor (*BBC*)
Christina Warr (*Course Assistant*)
B. G. Whatley (*BBC*)

The Open University Press
Walton Hall, Bletchley, Bucks

First Published 1971 ; 2nd edition 1973.
Copyright © 1973 The Open University

Designed by the Media Development Group of the Open University.

Printed in Great Britain by
Martin Cadbury Printing Group, Cheltenham and London

ISBN 0 335 02261 8

This text forms part of an Open University Second Level Course. The complete list of Units in the Course is given at the end of this text.

For general availability of this text and supporting material, please write to the Director of Marketing, The Open University, Walton Hall, Bletchley, Bucks.

Further information on Open University courses may be obtained from the Admissions Office, The Open University, PO Box 48, Bletchley, Bucks.

Unit 6   Oxidation States and the Typical Metals

Contents

Objectives

When you have completed this Unit you should be able to:

1   Define in your own words, recognize valid definitions of, or use in a correct context, the terms or expressions in Table A.

2   Derive the relative molar composition of a compound from its chemical formula, and then use a table of relative atomic masses to deduce the relative masses of the elements in the compound.

3   Derive the empirical formula of a compound from the relative masses of its constituent elements, given a table of relative atomic masses.

4   Balance simple chemical equations by inspection.

5   Calculate the changes in the masses of individual reactants and products in a chemical equation, given the chemical equation and a table of relative atomic masses.

6   Given an updated Mendeleev Periodic Table, describe the broad features of the periodicity in the formula of the highest normal oxides.

7   Show how the concept of fixed valency can link the different molecular compositions of different compounds of an element.

8   Assign oxidation numbers to elements in simple compounds or ions.

9   Use oxidation numbers to balance redox equations whose reactants and products are known.

10   Make an attempt to distinguish a typical element from a transition element when given the patterns of oxidation numbers in simple compounds.

11   Distinguish disproportionation reactions from other types of redox reaction, and redox reactions in general from other types of reaction.

12   Extract from the information given in Sections 6.2–6.3.1 inclusive, specific examples which demonstrate the principles and tendencies listed as 1–12 in Section 6.4.

13   Apply the principles and tendencies listed as 1–12 in Section 6.4 to appropriate examples which are not covered in Sections 6.2–6.3.1 inclusive.

Table A

List of scientific terms, concepts and principles used in Unit 6

Introduced in a previous Unit	Unit Section No.	Developed in this Unit	Page No.
	S100*		
		amphoteric oxide	19
acid	9.14	disproportionation	28
alkali	9.14	empirical formula	34
Avogadro's hypothesis	5.3.3	hydrate	21
base	9.10	hydrolysis	28
hydrogen bond	10.5	normal oxide	8
neutralization	9.8	oxidation	12
oxidation	8, App. 1	oxidation number	12
salt	9.8	peroxide	8
stoichiometry	6, App. 2	reduction	12
transition elements	8.3	redox reaction	14
typical elements	8.3		
	S25-		
Born-Haber cycle	5.8		
entropy	5.7		
Gibbs function	5.7		
kinetic stability	5.2		
thermodynamic stability	5.2		

* The Open University (1971) S100 *Science: A Foundation Course,* The Open University Press.

Study guide

The main text of this Unit falls into two parts. Sections 6.1–6.1.8 cover some of the patterns that stoichiometry reveals in the Periodic Table and discuss concepts, such as oxidation number, which chemists have introduced to express such patterns in a concise form.

The second part, Sections 6.2.1–6.3.1, uses the chemistry of the alkali metals, the alkaline earth metals and aluminium as a vehicle to demonstrate some important principles and tendencies, which are summarized in Section 6.4.

As we will be dealing with the subject of stoichiometry, which was introduced in the Foundation Course (S100, Unit 6, Appendix 2), we have included four objectives (Nos. 2–5), twelve self-assessment questions* (SAQs 1–12) and an appendix which revise this topic.

If you are not familiar with stoichiometric problems, Appendix 1 (White), Chemical formulae and stoichiometry, is certainly the most important part of this Unit for you.

The Home Experiments connected with this Unit can be spread over two working weeks as there is no Home Experiment for Unit 7. Unit 6 is long, but Unit 7 shorter. There is a lot of factual material in this Unit which you do not need to remember. Objectives 12 and 13 should help you to decide what to memorize.

* SAQs 1–12 are in Appendix 1 (pp. 34–6).

6.1 Introduction

In Units 3, 4 and 5 (besides introducing some elementary thermodynamics), we discussed the metals and, in particular, their structure and the relative ease with which we could extract them from the compounds that they form. For the most part, compounds were viewed only as potential sources of elements, and we paid very little attention to their chemistry. However, as it was mainly the study of the chemistry of compounds of the elements that first led to the discovery of the periodic law, it is perhaps in this chemistry that we shall find the clearest expression of chemical periodicity.

In this Unit, we begin our study of the chemistry of the typical elements and, at the same time, encounter some of the fundamental ideas that formed the foundation of the periodic law. The chemistry that we consider intensively will be mainly that of sodium, magnesium and aluminium, but we shall roam much more widely over the Periodic Table in studying one important chemical property. This property is the periodic variation in the formulae of certain compounds which we discuss from the beginning of the Unit.

The formulae of most compounds have been established by stoichiometry, which is the study of the amounts of substances that take part in chemical reactions or are combined together in chemical compounds. The subject of stoichiometry was discussed in S100, Unit 6, Appendix 2, and is revised in Appendix 1 of this Unit. There, you will find a dozen questions which will test your understanding of this very important subject and, in particular, your ability to achieve Objectives 2, 3, 4 and 5.

Once the atomic theory became widely accepted, the study of stoichiometry allowed the chemical formulae of compounds to be established. As we hope to convey in this Unit, such a step was essential to the development of concepts such as the periodic law, valency, and oxidation and reduction. In studying periodic variations we shall be looking especially at the third period of the Table, and we begin with the formulae of the halides of the first three elements.

6.1.1 The formulae of the sodium, magnesium and aluminium halides

Would you expect sodium, magnesium and aluminium to react with the halogens?

In the chloride series discussed in Unit 5, SAQ 9, sodium, magnesium and aluminium are, as usual, highly placed. This is because of the negative values of $\Delta G_f^\ominus$ for the chlorides and other halides, which show that the reactions between the metals and halogens have large equilibrium constants. Vigorous reactions do in fact occur on heating. The halides which are formed can also be made by heating the metals with the hydrogen halide gases, e.g.

$$Mg(s) + 2HCl(g) = MgCl_2(s) + H_2(g)$$

These halides are colourless solids, and we shall describe their chemistry later. For the present, we shall be concerned only with their formulae. These can be established by methods like that used in SAQ 10. For sodium, the formulae are of the type NaX (e.g. NaCl), while the magnesium halides have formulae MgX_2 and aluminium forms compounds of the type AlX_3.

6.1.2 Oxides of sodium, magnesium and aluminium

When heated in air, sodium, magnesium and aluminium combine vigorously with oxygen to form compounds with the formulae NaO, MgO and Al_2O_3.

Are you surprised?

Possibly! It seems odd that the sodium and magnesium oxides have the same type of formula, especially when we compare the smooth progression in halide formulae. However, when the three oxides are added to dilute acid, a significant difference is observed. All three dissolve, but whereas the magnesium and aluminium oxides undergo orthodox neutralization reactions like

$$Al_2O_3(s) + 6HNO_3(aq) = 2Al(NO_3)_3(aq) + 3H_2O$$

7

the sodium compound forms a solution from which oxygen gas is gradually evolved. At the same time it can be shown that the solution contains the compound hydrogen peroxide H_2O_2:

$$2NaO(s) + 2HCl(aq) = 2NaCl(aq) + H_2O_2(aq) \qquad (1)$$

and that the oxygen gas is produced by the slow decomposition of the hydrogen peroxide:

$$H_2O_2(aq) = H_2O + \tfrac{1}{2}O_2(g) \qquad (2)$$

What is the final overall reaction?

Add equations 1 and 2 to get

$$2NaO(s) + 2HCl(aq) = 2NaCl(aq) + H_2O + \tfrac{1}{2}O_2(g)$$

(a)

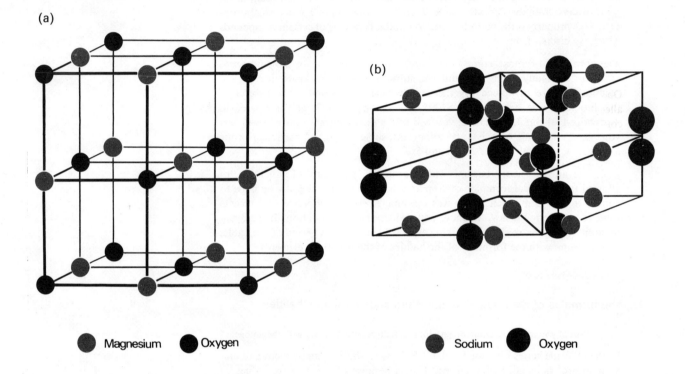

| Magnesium | Oxygen | | Sodium | Oxygen |

This difference in chemical behaviour is matched by a difference in crystal structure which is shown in Figure 1. In MgO, as in most oxides, isolated oxygen atoms can be picked out, and in the rock salt structure of MgO each one is surrounded by six magnesium neighbours. However, in NaO the oxygens are grouped in pairs with an internuclear distance not very much greater than in oxygen gas. Consequently, in an ionic bonding scheme, the anion in MgO is the conventional O^{2-} ion, but in NaO it is written as O_2^{2-}. The ion, O_2^{2-}, is called the peroxide anion and its relation to the compound hydrogen peroxide,

$$\begin{array}{c} \diagup O\!-\!O\diagdown \\ H \qquad\qquad H \end{array}$$

Figure 1 Structure of (a) MgO and (b) NaO showing the grouping of the oxygens in the latter.

is the same as that of the sulphate ion, SO_4^{2-}, to sulphuric acid, H_2SO_4. Therefore, NaO is usually written Na_2O_2 and called sodium peroxide.

peroxide

If sodium peroxide is strongly heated in a vacuum, oxygen gas is evolved, and if this is pumped away as it is formed, Na_2O is left:

$$Na_2O_2(s) = Na_2O(s) + \tfrac{1}{2}O_2(g)$$

Na_2O has a *normal oxide* structure* which we shall discuss in detail in Unit 7.

normal oxide

* We define a normal oxide as one in which the oxygen atoms are isolated from one another. Such compounds usually undergo orthodox neutralization reactions to give a salt plus water only.

We see, therefore, that if we divest ourselves of exceptional cases, a progressive increase in oxygen content is observed in the normal oxide series Na_2O, MgO and Al_2O_3.

6.1.3   Periodic variations in the formulae of compounds

As we move from neon, through sodium and magnesium to aluminium oxide, the number of oxygen atoms per atom of metal in the chemical formulae of the normal oxides increases by one-half for each step.

> Using elementary bonding theories can you account for this change?

If you assume that the compounds contain ions with the stable electronic configuration of a noble gas, then they are composed of the ions Na^+, Mg^{2+}, Al^{3+} and O^{2-}. It follows that if the compounds are to be electrically neutral, the formulae of the oxides must be Na_2O, MgO and Al_2O_3. This is the observed result. A similar argument could be applied to the halides NaX, MgX_2 and AlX_3.

However, the regular change in the formulae of the normal oxides does not stop at aluminium; the oxide of the next element, silicon, has the expected formula SiO_2. Unlike the elements from sodium to silicon, phosphorus and sulphur and chlorine form more than one normal oxide, but if we look for the normal oxides with the largest number of oxygen atoms per atom of the other elements, we find that their formulae are the 'expected' ones.

> What are the 'expected' ones?

They are P_2O_5, SO_3 and Cl_2O_7. These, and other normal oxides with the greatest oxygen content for a given element, we shall call the *highest normal oxide* of that element.

Now, the steady change in stoichiometry is not confined to the highest oxides, although it is in these compounds that it is most clearly seen. In Table 1, the formulae of the highest normal oxides, fluorides and chlorides of the elements of the third period are given. For the fluorides, the number of fluorine atoms per atom of the element is equal to the group number, except at chlorine where the pattern is broken and it is two less. For the chloride sequence, the exceptions are sulphur and chlorine.

Table 1   Highest normal oxides, fluorides and chlorides of the elements of the third period

0	1	2	3	4	5	6	7
Ne	Na	Mg	Al	Si	P	S	Cl
—	Na_2O	MgO	Al_2O_3	SiO_2	P_2O_5	SO_3	Cl_2O_7
—	NaF	MgF_2	AlF_3	SiF_4	PF_5	SF_6	ClF_5
—	$NaCl$	$MgCl_2$	$AlCl_3$	$SiCl$	PCl_5	SCl_4	Cl_2

These periodic changes in stoichiometry show little respect for our elementary bonding theories. Take, for example, the fluorides. The sodium, magnesium and aluminium compounds are involatile solids with, as you will see later, typical lattices of the ionic type. However, at room temperature, silicon tetrafluoride, phosphorus pentafluoride and sulphur hexafluoride are gases. All three vapours contain discrete molecules which can therefore only be regarded as covalent. Thus the regular change in stoichiometry transcends what is, by our experimental criteria, a sharp transition from ionic to covalent fluorides between aluminium and silicon.

Figure 2 Shapes of phosphorus pentafluoride and sulphur hexafluoride molecules in the gas phase.

Notice especially that we cannot describe the gaseous compounds PF_5 and SF_6 in terms of Lewis structures in which each atom has a noble gas structure. The geometry of the two molecules is shown in Figure 2. Phosphorus, which has five outer electrons, forms five bonds and sulphur, with six outer electrons, forms six. If we assign two electrons to each bond, phosphorus in PF_5 contains ten outer electrons and sulphur in SF_6 has twelve. This contrasts with phosphorus and sulphur molecules such as PF_3 and SCl_2 in which the assignment of two electrons to each bond gives phosphorus and sulphur a configuration that you are used to: the eight outer electrons associated with noble gas configurations.

Whatever the theoretical problems associated with the regular variations in stoichiometry that we have discussed in this Section, there is no doubt that, historically, the variations were of the greatest importance in attempts to bring order to the chemistry of these elements. This is especially true of the periodic changes in the formulae of the highest normal oxide, for it was perhaps upon these changes more than anything else that Mendeleev based his Periodic Table. With no structural techniques to help him, he distinguished peroxides from normal or, as he called them, saline oxides, by the chemical behaviour with acids described for Na_2O_2. As he remarked in the English translation of his book *The Principles of Chemistry** published in 1905: 'A slight acquaintance with bases, acids and salts, and with peroxide of hydrogen, removes all doubt as to whether a given oxide or its hydrate should be referred to the class of saline oxides, or of peroxides.'

In Figure 3, the form of the updated Mendeleev Table is presented, but the symbols of the elements have been replaced by the formulae of the highest known normal oxides and fluorides.

The periodic pattern is clearly seen. For the oxides, the only failures are found at oxygen, fluorine, the three Group Ib metals, the Group VIII triads (except for ruthenium and osmium), three lanthanides, and at xenon and radon where recent research has established the existence of XeO_4 and some poorly characterized oxygen compounds of radon.

Figure 3 The Mendeleev periodicity in the formulae of highest normal oxides and fluorides.

0	I		II		III		IV		V		VI		VII		VIII		
	a	b	a	b	a	b	a	b	a	b	a	b	a	b			
He	Li_2O		BeO		B_2O_3		CO_2		N_2O_5		OO_2		F_2O				
	LiF		BeF_2		BF_3		CF_4		NF_3		OF_2		F_2				
Ne	Na_2O		MgO		Al_2O_3		SiO_2		P_2O_5		SO_3		Cl_2O_7				
	NaF		MgF_2		AlF_3		SiF_4		PF_5		SF_6		ClF_5				
Ar	K_2O		CaO		Sc_2O_3		TiO_2		V_2O_5		CrO_3		Mn_2O_7		Fe_2O_3	Co_2O_3	NiO_2
	KF		CaF_2		ScF_3		TiF_4		VF_5		CrF_6		MnF_4		FeF_3	CoF_3	NiF_3
		CuO		ZnO		Ga_2O_3		GeO_2		As_2O_5		SeO_3		?			
		CuF_2		ZnF_2		GaF_3		GeF_4		AsF_5		SeF_5		BrF_5			
Kr	Rb_2O		SrO		Y_2O_3		ZrO_2		Nb_2O_5		MoO_3		Tc_2O_7		RuO_4	Rh_2O_3	PdO_2
KrF_2	RbF		SrF_2		YF_3		ZrF_4		NbF_5		MoF_6		TcF_6		RuF_6	RhF_6	PdF_6
		AgO		CdO		In_2O_3		SnO_2		Sb_2O_5		TeO_3		I_2O_7			
		AgF_2		CdF_2		InF_3		SnF_4		SbF_5		TeF_6		IF_7			
XeO_4	Cs_2O		BaO		Ln_2O_3*		HfO_2		Ta_2O_5		WO_3		Re_2O_7		OsO_4	IrO_2	PtO_2
XeF_6	CsF		BaF_2		LnF_3**		HfF_4		TaF_5		WF_6		ReF_7		OsF_6	IrF_6	PtF_6
		Au_2O_3		HgO		Tl_2O_3		PbO_2		Bi_2O_5		PoO_3		?			
		AuF_3		HgF_2		TlF_3		PbF_4		BiF_5		PoF_6		?			
?	Fr_2O		RaO		Ac_2O_3		ThO_2		Pa_2O_5		UO_3						
RnF_6	FrF		RaF_2		AcF_3		ThF_4		PaF_5		UF_6						

* Cerium, praseodymium and terbium form MO_2.

** Cerium, praseodymium and terbium form MF_4.

* D. I. Mendeleev (1905) *The Principles of Chemistry*, 3rd ed., Longmans.

More than anything else, it was the periodic pattern in the formulae of the highest normal oxides which gave Mendeleev's table its characteristic if somewhat contrived appearance. Look at the row that begins with copper. This row displays something close to the correct periodicity in the formulae of the highest normal oxide only because three elements which do not fit into the pattern, iron, cobalt and nickel, have been left hanging in the air in Group VIII of the preceding row. Likewise, the periodicity in the row beginning with caesium (we ignore xenon as it had not been discovered when Mendeleev proposed his law) has been maintained only by packing the 15 lanthanide elements into one space after barium.

The most contrived features of the Mendeleev-style table were in large part designed to maintain the periodicity in the formulae of the highest normal oxides.

Now do SAQs 13 and 14 (on p. 37).

6.1.4 Valency

We have now looked carefully at the sequence of the highest oxides as we move across the third period, and also at those in the highest fluorides and chlorides. What we have not done is to look at the relations between the individual sequences. Look again at the formulae in Table 1. If the halides of sulphur and chlorine are ignored, it looks as though fluorine and chlorine atoms combine with equal numbers of other atoms, but oxygen atoms possess twice the combining power of both.

An early way of expressing this stoichiometric relationship was to say that each element could form only a fixed number of bonds, or had a fixed *valency*. Such an idea can account for the sequences in Table 1.

For example, if oxygen is divalent, and the halogens are for the most part univalent, gaseous molecules of SF_6 and SO_3 can both be written as compounds of hexavalent sulphur. This is shown in Figure 4 which should be considered as two-dimensional.

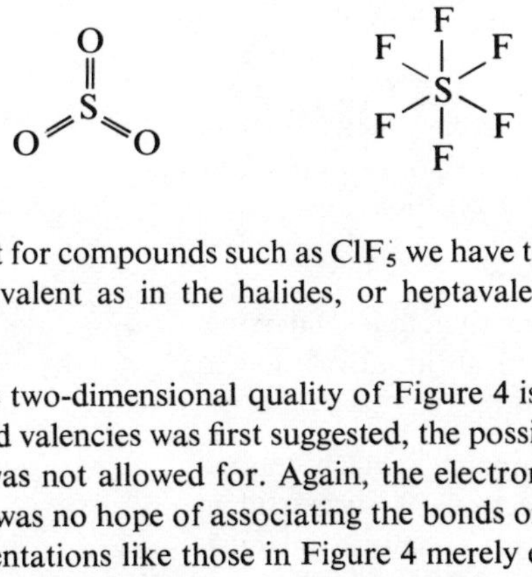

Figure 4 Two compounds of hexavalent sulphur.

Of course, to account for compounds such as ClF_5 we have to allow that chlorine can not only be univalent as in the halides, or heptavalent as in Cl_2O_7, but pentavalent too.

The emphasis on the two-dimensional quality of Figure 4 is important because, when the idea of fixed valencies was first suggested, the possibility that molecules might have shapes was not allowed for. Again, the electron had not even been discovered, so there was no hope of associating the bonds or links with numbers of electrons. Representations like those in Figure 4 merely expressed a relationship between the formulae of different compounds. However, this was a very important advance. Consider, for example, the enormous number of compounds of carbon and hydrogen mentioned in S100, Unit 10. Nearly all of these have a common feature: they can be described by diagrams in which each carbon atom forms four bonds, and each hydrogen atom one.

To see if you appreciate this, try SAQ 15 (on p. 37).

Once we move away from organic chemistry, the idea of fixed valencies runs into difficulties. For the compounds in Table 1, for example, to account for ClF_5, we have to allow that chlorine can not only be univalent as in the halides, or heptavalent as in Cl_2O_7, but pentavalent too. Nevertheless, the idea retains considerable value, because many elements show a *preference* for some valencies rather than others. Indeed, this accounts for the relationship between the different sequences in Table 1.

However, as structural information grew, it became clear that formulae such as those in Figure 4 ought to be given three-dimensional qualities. What was more, this information threw the very idea of valency as a unifying principle into disfavour – for example, it seemed inconsistent to write NaCl as Na-Cl when each sodium in a salt crystal was surrounded by six chlorines. Attempts were made to find a different concept that, without carrying structural implications, expressed in a concise way the relationship between sequences like those in Table 1 and, at the same time, retained the advantages of the idea of preferred valencies. One of the commonest now used is closely tied to the twin concepts of oxidation and reduction.

6.1.5 Oxidation and reduction

In Unit 3, we used the definitions of oxidation and reduction that were first introduced in S100, Unit 8; oxidation was defined as loss of electrons, and reduction as gain of electrons. In the next Section we introduce a new definition of oxidation and reduction, and the need for this development is best demonstrated by beginning with the earliest meaning of oxidation: the reaction of an element or compound with oxygen. With this definition, the idea of oxidation seems simple. When magnesium is heated in oxygen, magnesium oxide is formed

$$Mg(s) + \tfrac{1}{2}O_2(g) \longrightarrow MgO(s) \tag{3}$$

and the magnesium has been oxidized. However, this is not the only way of converting magnesium into its oxide. We could heat the metal in chlorine to get $MgCl_2$, dissolve the chloride in water, add sodium hydroxide to precipitate the insoluble compound $Mg(OH)_2$ and filter the hydroxide. When heated, the hydroxide loses water leaving the oxide.

$$Mg(s) + Cl_2(g) = MgCl_2(s) \tag{4}$$

$$MgCl_2(s) = MgCl_2(aq) \tag{5}$$

$$MgCl_2(aq) + 2NaOH(aq) = Mg(OH)_2(s) + 2NaCl(aq) \tag{6}$$

$$Mg(OH)_2(s) = MgO(s) + H_2O(g) \tag{7}$$

Again, the magnesium has been oxidized, and the question arises: in which one of the four steps has oxidation occurred? Chemists have always felt that the first step was the most reasonable choice, although you may think that a case could be made for the third.

Yet another way of obtaining the oxide would be to dissolve magnesium in dilute HCl, and treat the resulting solution of $MgCl_2$ in the ways implied by equations 6 and 7. Here, oxidation was felt to occur in the reaction

$$Mg(s) + 2H^+(aq) = Mg^{2+}(aq) + H_2(g)$$

Thus we have a situation in which the conversion of magnesium to a solid oxide, a solid chloride, $MgCl_2$, or an aqueous ion, Mg^{2+}, are all intuitively felt to involve oxidation of magnesium.

Now, if we suppose MgO and $MgCl_2$ to be ionic compounds which contain Mg^{2+} ions, our definition of oxidation as loss of electrons is successful in labelling these three conversions as oxidation. However, as we shall see in Unit 7, there is evidence that we are not entirely correct in describing $MgCl_2$ as an ionic compound. What is more, reactions such as

$$C(s) + \tfrac{1}{2}O_2(g) = CO(g)$$

which describe the formation of covalent compounds by oxidation, cannot be discussed in terms of loss or gain of electrons. To bring this kind of reaction, and reactions such as 3 and 4 under the single heading of oxidation reactions, we need a new definition. This definition involves the concept of oxidation number.

6.1.6 Oxidation number

Oxidation numbers, also called oxidation states, often provide a useful method of classifying the ions and compounds that an element forms. This is particularly

oxidation number

12

true in the chemistry of metals. We shall first simply state the rules used to assign them, and then get you to practise working them out.

It is not possible to obtain a perfect scheme for the comprehensive assignment of oxidation numbers, but the following should suffice for any cases that you encounter in this Course.

1   The oxidation number of a monatomic ion in solution is equal to the charge on the ion. Thus, the oxidation numbers of iron and chlorine in Fe^{2+} (aq) and Cl^-(aq) are $+2$ and -1 respectively.

2   The oxidation number of an atom in its elemental form is zero. Thus the values for aluminium and bromine atoms in Al(s) and Br_2(l) respectively are zero.

3   The oxidation number of oxygen in compounds is -2, except in compounds such as peroxides where oxygen atoms are linked together. Thus in CO_2, the oxygen atom has an oxidation number of -2.

4   The oxidation number of hydrogen in compounds is taken to be $+1$, except in metallic hydrides such as NaH where it is -1. Thus in HCl, the oxidation number of hydrogen is $+1$.

5   The oxidation number of halogen atoms in compounds is -1, except in the compounds that they form with oxygen and with one another. Thus in HCl the oxidation number of chlorine is -1.

6   The oxidation number of a metal in a metallic salt is positive, and equal to the number of hydrogen atoms which each metal atom has displaced from the acid from which the salt is derived. Thus silver sulphate, Ag_2SO_4, is derived from sulphuric acid, H_2SO_4. Two hydrogens have been replaced by two silvers, so the oxidation number of silver is $+1$.

7   The sum of the oxidation numbers of the atoms in a compound or ion is equal to the charge on that compound or ion. Thus HCl has no charge, and the oxidation numbers of H and Cl are $+1$ and -1.

8   When oxidation numbers are assigned in this way, *an atom is said to have been oxidized in a reaction when its oxidation number has increased. Reduction is defined as a decrease in oxidation number.*

Before going on, test your mastery of these rules by doing SAQ 16 (on p. 37).

The examples in the SAQs are fairly straightforward, but even so, you may feel that the eight rules contain too many exceptions and equivocations to make the definition of oxidation or of oxidation number watertight. You are quite right. Oxidation numbers are not fundamental quantities and you should note especially that they cannot in general be taken as an indication of the charge that an atom carries in a compound. There is, for example, no sense in which the sulphur atom in gaseous SO_2 has a charge of $+4$. In the main, oxidation numbers represent an artificial attempt by chemists to classify compounds of a particular element in accordance with intuitive feelings generated by a wide experience of chemical reactions, e.g. the feeling that Mg^{2+} (aq), $MgCl_2$ and MgO have something in common with each other that they do not share with magnesium metal. Again, the mere act of dissolving SO_2 in water gives H_2SO_3 which is easily neutralized with NaOH to give Na_2SO_3. Likewise, SO_3 gives H_2SO_4 in water which yields Na_2SO_4 with NaOH. Chemists feel that SO_2, H_2SO_3 and $NaSO_3$ have something in common which they do not share with SO_3, H_2SO_4 and Na_2SO_4.

Definitions rooted in intuitive feelings of this kind cannot be expected to be watertight, and oxidation numbers are no exception. One might have hoped, for example, that the carbon atoms in CH_4, C_2H_6 and C_3H_8 with their similar valencies and similar reactions would all be in the same oxidation state. However, we must either introduce some new rule so that we can give different carbon atoms different oxidation states or allow their oxidation numbers to be -4, -3 and $-2\frac{2}{3}$ respectively! On the other hand, diagrams of the type shown in Figure 4 would clearly establish a relationship between CH_4, C_2H_6 and C_3H_8 through the tetravalency of carbon. Thus in some areas of chemistry, especially organic chemistry, valency affords a more satisfactory way of classifying compounds than does oxidation number. The latter is most useful for compounds of metals.

In spite of the kind of example that we have just mentioned, oxidation numbers are at present the most watertight way we have of giving expression to what chemists mean by oxidation and reduction. For example, they overcome the difficulty mentioned in Section 6.1.5.

Notice that, taken together, rules 7 and 8 imply that oxidation cannot occur without reduction. Thus, in the combustion of magnesium, the oxidation number of the metal changes from zero to +2, while that of oxygen falls from zero to −2; the metal has been oxidized and the oxygen reduced. This should become even clearer in the next Section where we discuss a use of oxidation number that is less important than the classification role but easier to describe in a precise way.

6.1.7   The balancing of oxidation-reduction equations

Rules 7 and 8 have a second important consequence. As the sum of the oxidation numbers in a compound or ion is equal to its charge (and charge is conserved in a chemical reaction), the sum of the changes in oxidation numbers in a balanced chemical equation must be zero. This deduction can be used to balance equations. Until now, you have encountered equations which can be balanced by careful inspection. This was true of those in SAQ 7. However, oxidation-reduction reactions, also called *redox* reactions, are often hard to balance by inspection. Oxidation numbers make the job easier.

redox reaction

Consider, for example, the reaction between an acid solution of potassium permanganate, $KMnO_4$, and hydrogen sulphide gas. The purple solution of the permanganate is decolorized, and a yellow precipitate of sulphur appears. The solution left contains the nearly colourless Mn^{2+} (aq) ion. The initial solution of $KMnO_4$ contained the ions K^+(aq) and MnO_4^-(aq), and as K^+(aq) is not changed, we can leave it out and, to begin with, write the equation

$$MnO_4^- (aq) + H_2S(g) = Mn^{2+} (aq) + S(s) \tag{8}$$

This equation summarizes the observations made on the reaction. We now have to balance it.

1   The first thing to notice is that oxygen appears on the left-hand side of equation 8 but not on the right. The oxygen in MnO_4^- must have been transformed into something and, since no oxygen gas was observed during the reaction, it seems reasonable to suppose that it was converted into water. The production or consumption of water in redox reactions in aqueous solution is very common. We therefore rewrite the equation:

$$MnO_4^- (aq) + H_2S(g) = Mn^{2+} (aq) + S(s) + H_2O$$

2   Write appropriate oxidation numbers over each atom in the above equation.

You should have got

$$\overset{+7\ -2}{MnO_4^-} + \overset{+1\ -2}{H_2S} = \overset{+2}{Mn^{2+}} + \overset{0}{S} + \overset{+1\ -2}{H_2O}$$

3   Which atoms have been reduced and which oxidized?

Sulphur has been oxidized (−2 to 0) and manganese has been reduced (+7 to +2). The oxidation numbers of other atoms do not change.

4   How many moles of H_2S must react with one mole of MnO_4^-, if the sum of the changes in oxidation number is to be zero?

Since the oxidation number of manganese has been reduced by five and that of sulphur has been raised by two, $2\frac{1}{2}$ moles of H_2S must react with each mole of MnO_4^-. Converting to the lowest possible whole numbers, 2 moles of MnO_4^- react with 5 moles of H_2S. We now write the equation with this stoichiometry, and ensure that the manganese and sulphur atoms are conserved (i.e. their numbers are the same on both sides of the equation):

$$2MnO_4^- + 5H_2S = 2Mn^{2+} + 5S + H_2O$$

5   We now ensure conservation of oxygen atoms. There are eight of these on the left, so eight molecules of water are produced on the right:

$$2MnO_4^- + 5H_2S = 2Mn^{2+} + 5S + 8H_2O$$

14

6 Our final step is to ensure conservation of hydrogen atoms. There are 16 on the right and only 10 on the left. In neutral or acid aqueous solutions, we assume that the six needed on the left must have come from aqueous hydrogen ions. The final equation is then:

$$2MnO_4^-(aq) + 5H_2S(g) + 6H^+(aq) = 2Mn^{2+}(aq) + 5S(s) + 8H_2O$$

If we have performed our operations correctly, *the charges should balance*. This is a very useful check. As you can see, the charges sum to $+4$ on each side.

Once you have worked through a few examples, such as those in SAQ 18 (on p. 37), you should be able to perform this kind of exercise quite quickly.

6.1.8 Oxidation states of the typical elements

We have now transformed our examination of the stoichiometric variations that occur in the formulae of oxides and halides into a study of the changes in a single factor, namely, the oxidation state. For example, as we move across the third period from sodium to chlorine, the largest occurring oxidation number for the elements increases step by step from $+1$ to $+7$. Indeed, at room temperature, the compounds of sodium, magnesium and aluminium are almost entirely confined to the oxidation states $+1$, $+2$ and $+3$ respectively.

The other elements in the period show more variety in their oxidation states, but the variety conforms, for the most part, to a pattern which distinguishes the typical elements from other elements in the Periodic Table. If we ignore the elemental forms, and *the compounds in which the atoms of the typical elements are bonded to one another* (e.g. H_2O_2, Si_3H_8), the oxidation states of the typical elements usually, but not always, differ by multiples of two. Consider, for example, the different forms in which chlorine is found in aqueous solution. These are $Cl^-(aq)$, $ClO^-(aq)$ $ClO_2^-(aq)$, $ClO_3^-(aq)$ and $ClO_4^-(aq)$ (see Unit 11).

What are the oxidation numbers of chlorine in these ions?

-1, $+1$, $+3$, $+5$ and $+7$ respectively; they increase in steps of two. Again, let us look at the example of the phosphorus chlorides. There are two which exist at room temperature. They are PCl_3 and PCl_5 in which the oxidation state of phosphorus is $+3$ and $+5$. PCl_4 is not known.

Would PCl_4 contain an odd or even number of electrons?

It would contain an odd number. In fact, if we assume Lewis-type two-electron bonds, the outer shell of phosphorus contains nine electrons, one of which is not paired with another. This kind of arrangement is not usually favoured by the typical elements—if you like you can imagine the odd electron as representing unrealized bonding capacity—molecules or ions containing pairs of electrons are usually formed. In Table 2, we give the known simple fluorides of the third period and, as you can see, they all conform to the pattern that we have described.

Table 2 Simple fluorides of the third period which have been characterized at room temperature

					SF$_2$	ClF
				PF$_3$	SF$_4$	ClF$_3$
NaF	MgF$_2$	AlF$_3$	SiF$_4$	PF$_5$	SF$_6$	ClF$_5$

(Compounds such as S_2F_{10}, in which two third period atoms are bonded together, have been ignored.)

The situation is quite different in the case of the transition elements. We can compare phosphorus with vanadium. Both have the outer electron configurations containing five electrons; phosphorus with $3s^23p^3$ and vanadium with $3d^34s^2$. Both form chlorides MCl_3 and MCl_5 but vanadium also forms a tetrachloride, which is a red liquid made up of discrete VCl_4 molecules which contain an odd electron. It seems, then, that for some reason, the electrons in the compounds of the transition elements are less strongly inclined to be associated in pairs than the electrons in the compounds of the typical elements. One consequence of this is that the known oxidation numbers of the transition elements very often differ by units of only one. Thus chromium forms the fluorides CrF_2, CrF_3, CrF_4, CrF_5 and CrF_6. Compare the fluorides of sulphur in Table 2.

As you will learn in a third level Course, the *unpaired* electrons in transition metal compounds have a strong influence on physical and chemical properties of the different oxidation states. In addition, the different oxidation states of transition metals often have beautiful colourations. and you will encounter some of these in this Unit's Home Experiment.

Now try SAQ 17 (on p. 37).

6.2   Sodium, magnesium and aluminium chemistry

Most of the remainder of this Unit is devoted to the study of the compounds of sodium, magnesium and aluminium. The prime purpose, however, is not a comprehensive survey of the chemistry of these three elements; indeed, many important details of that chemistry have been omitted. It is rather to use selected compounds of sodium, magnesium and aluminium as a vehicle which introduces you to important general principles and tendencies which can then be applied to wider areas of inorganic chemistry. Some of the more important of these are summarized in Section 6.4, but we suggest that you read the text through quickly before examining them. You can then make a more careful study of the text while keeping the important principles in mind, and paying attention to Objectives 12 and 13.

Another feature of the chemistry of the three elements is that two of them, sodium and magnesium, behave very much like the other elements in the same group of the Periodic Table. This point is discussed in more detail in Section 6.3; it means that in studying the chemistry of sodium and magnesium we come close to a study of all the alkali and alkaline earth metals. Nevertheless, some important variations do occur within these two families, and these are discussed in Sections 6.3 – 6.3.1.

6.2.1   Sodium, magnesium and aluminium metals

S100, Units 8 and 9, and Unit 3 of this Course contain a considerable amount of information on sodium, magnesium and aluminium metals.

What is common to the three modern methods used to obtain these metals?

As with all metals at the top of the various series discussed in Unit 5, the key step in the extraction process usually involves electrolysis. In the case of sodium (Unit 3) and magnesium (S100, Unit 9), it is electrolysis of the fused chlorides; the earliest methods used by Davy (Unit 3, Appendix 1) are not applied on an industrial scale. The extraction of aluminium was described in Unit 3, Appendix 1 and in the TV programme for that Unit.

At room temperature, sodium combines vigorously with the oxygen of the air, and dissolves in water to form the alkaline solution of the hydroxide, and hydrogen gas. With magnesium and aluminium, these reactions are barely noticeable. Magnesium and aluminium hydroxides are insoluble and the expected reactions with water are

$$Mg(s) + 2H_2O = Mg(OH)_2(s) + H_2(g) \quad \Delta G_m^\ominus = -359.5 \text{ kJ mol}^{-1}$$

$$Al(s) + 3H_2O = Al(OH)_3(s) + \tfrac{3}{2}H_2(g) \quad \Delta G_m^\ominus = -432.1 \text{ kJ mol}^{-1}$$

What do you deduce from this?

The signs and magnitudes of $\Delta G_m^\ominus$ show that the reactions have large equilibrium constants. This means that their reluctance to proceed is caused by a slow rate of reaction.

Can you suggest a reason for this?

One possible reason is supplied by the insolubility of the oxides or hydrated oxides that are produced. A thin insoluble film covers the metal and slows the reaction down. With both magnesium and aluminium (Unit 3, TV programme), the evolution of hydrogen in water is brisk if the coherence of the oxide film is

16

destroyed by amalgamation. Amalgamation also causes the metals to corrode quickly in air.

The oxide film even protects aluminium in nitric acid (Unit 3, TV programme) although the metal dissolves rapidly in hydrochloric and sulphuric acids. Nitric acid is a strong oxidizing agent and it is as if, in this instance, the acid qualities of nitric acid take second place to its oxidizing properties which tend to coat the metal with oxide.

6.2.2 Oxygen compounds

Sodium combines readily with oxygen at room temperature. Magnesium and aluminium are *kinetically* stable with respect to oxygen but if they are heated sufficiently strongly, the protective oxide film is disrupted, and the metals burn brilliantly.

> Can you suggest a reason why heating might disrupt the oxide film?

The oxide films afford protection to magnesium and aluminium because it so happens that when a certain amount of the two metals is oxidized, the volume of the oxide product is *almost the same as that of the original metal*. Heating could increase the mobility of the underlying metal by melting it, or cause the metal and oxide covering to expand at different rates. The products of the combustion of magnesium and aluminium are MgO and Al_2O_3 respectively.

> Is there a difference from sodium here?

Remember Section 6.1.2. The product of sodium combustion is mainly the peroxide, Na_2O_2. Magnesium and aluminium peroxides are not produced in this way; indeed, both are unknown in the pure state.

> Can you relate the absence of MgO_2 in the combustion products of magnesium to a particular reaction?

It may well be that at the high temperatures created by magnesium combustion, magnesium peroxide is unstable with respect to the reaction

$$MgO_2(s) = MgO(s) + \tfrac{1}{2}O_2(g)$$

As stated above, magnesium peroxide has not yet been obtained in the pure state; attempts to make it, even at room temperature, usually yield a product containing magnesium oxide, presumably because of the instability of the peroxide at normal temperatures with respect to the reaction given above. This contrasts with the stability of sodium peroxide in an analogous situation for which thermodynamic data are given in Table 3; note the positive value of $\Delta G_m^\ominus$.

However, the decomposition is endothermic, so the equilibrium constant increases with temperature. Although the peroxide is stable at 25 °C, a sufficient increase in temperature will lower $\Delta G_m^\ominus$, increase the equilibrium constant and make it unstable. A similar problem is discussed in more detail in Section 6.3.1.

Table 3 Thermodynamic properties of the decomposition of Na_2O_2 at 25°C

Reaction	$\Delta G_m^\ominus/$ kJ mol^{-1}	$\Delta H_m^\ominus/$ kJ mol^{-1}	$\Delta S_m^\ominus/$ J K^{-1} mol^{-1}
$Na_2O_2(s) = Na_2O(s) + \tfrac{1}{2}O_2(g)$	56·3	88·7	108·4

The non-existence of pure magnesium peroxide, and of aluminium peroxide, $Al_2(O_2)_3$, at room temperature suggests that $\Delta G_m^\ominus$ for a decomposition reaction like

$$Al_2(O_2)_3 = Al_2O_3 + \tfrac{3}{2}O_2$$

is negative at 25 °C.

To relate the non-existence of a particular compound to a particular decomposition reaction is a vital first step in any effort to explain its instability. Indeed the words 'stability' or 'instability' have no meaning unless this reaction is specified. This is an important part of the theme of this Unit's TV programme.

6.2.3 Hydroxides

The oxides Na_2O, MgO and Al_2O_3 all combine with water to form hydroxides. NaOH is extremely soluble in water, and the solution is a strong alkali because it contains a high concentration of hydroxide ions.

The solubility of sodium hydroxide is a very unusual property. Many other metals form hydroxides, but very few of them are appreciably soluble in water or, therefore, form alkaline solutions. Indeed, if hydroxide ions are added to the solutions of most metallic salts, insoluble hydroxides or hydrated oxides are precipitated.

If we consider solutions containing appreciable concentrations of the monatomic metallic cations known to exist in aqueous solutions, then in the whole of the Periodic Table, the only cations which are not precipitated by the addition of hydroxide ions are the alkali metal ions of the type M^+, Sr^{2+}, Ba^{2+}, Ra^{2+} and Tl^+. Even among these, lithium and strontium hydroxides are only slightly soluble.

The exceptions do not include the hydroxides of magnesium and aluminium. These can be precipitated from solutions of their salts by adding hydroxide ions.

Can you recall an important application of this reaction for magnesium?

Hydroxide precipitation is used to extract magnesium from sea water (S100, Unit 9).

Although magnesium hydroxide can be precipitated in this way, it is not so insoluble that it fails to give alkaline properties when added to water. A suspension of $Mg(OH)_2$ turns red litmus blue. This is not true of the corresponding aluminium compound which shows other important differences from magnesium hydroxide.

If a little sodium hydroxide is added to a solution of an aluminium salt, a white precipitate appears. If more sodium hydroxide is added, the precipitate redissolves. The addition of ammonia to a solution containing Al^{3+} ions also gives the white precipitate, but the latter is *not* soluble in an excess of ammonia.

The solid that is precipitated is a hydrated aluminium oxide. Depending upon the conditions of precipitation, aluminium can form several different hydrated oxides.

These differ both in structure and in the amount of water that they contain, but chemically they behave similarly, and here we shall describe them all by the blanket term 'aluminium hydroxide', '$Al(OH)_3$'.

If the hydroxide is added to water, the alkaline qualities of the liquid are not enhanced. This is because the solubility of aluminium hydroxide is very small, and in particular, the dissociation,

$$Al(OH)_3(s) = Al^{3+}(aq) + 3OH^-(aq)$$

hardly occurs at all in pure water. However, if an acid is added, the equilibrium is perturbed by the removal of hydroxide ions, and the solid can dissolve. In other words, aluminium hydroxide reacts with hydrogen ions:

$$Al(OH)_3 + 3H^+(aq) = Al^{3+}(aq) + 3H_2O$$

In this reaction therefore, $Al(OH)_3$ acts as a base.

However, the willingness of the hydroxide to dissolve in caustic alkalis suggests that, in a sense, it can act as an acid too by combining with hydroxide ions. This is so. Water in which the hydroxide is suspended has neither acid nor alkaline qualities, but if $[OH^-]$ is increased sufficiently, the equilibrium

$$Al(OH)_3(s) + OH^-(aq) = Al(OH)_4^-(aq)$$

is driven over to the right, and an *aluminate* anion is formed.

In this reaction therefore, $Al(OH)_3$ acts as an acid.

Why does the hydroxide not dissolve in ammonia, which is an alkali?

Ammonia is a weak alkali, and supplies only a low concentration of hydroxide ions (S100, Unit 9). This concentration is insufficient to displace the equilibrium enough to dissolve the hydroxide. Dissolution occurs only in alkalis stronger than ammonia.

In the sense that it can neutralize both acids and alkalis, aluminium hydroxide can act as either a base or an acid. Oxides which, whether by themselves or as their hydroxide derivatives, can play this dual role, are said to be *amphoteric*. In this respect, Al_2O_3 contrasts with both MgO and Na_2O which are purely basic oxides.

amphoteric oxide

This contrast can be related to a tendency which persists right across the group.

What can you say about the basic character of the oxides or hydroxides as one moves from sodium to aluminium?

By basic character, we mean the tendency of an oxide or hydroxide to form hydroxide ions in aqueous solution. As we move from sodium to aluminium the solubility* of the oxides and hydroxides decreases (the oxides form hydroxides with water) and the concentration of hydroxide ions in a saturated solution falls off. Thus basic character decreases.

At the same time, acid character increases, as shown by the solubility of aluminium hydroxide in caustic alkalis. This progression is apparent right across the period as we move from higher oxide to higher oxide. For example, SiO_2 does not react readily with either acids or alkalis, but if anything it dissolves more willingly in alkalis giving silicates such as Na_2SiO_3. Phosphorus pentoxide, P_2O_5, is an entirely acidic oxide, although H_3PO_4 (which is formed when it is added to water) is not a very strong acid. This contrasts with sulphuric and perchloric acids which can be obtained by dissolving the parent oxides SO_3 and Cl_2O_7 in water, and are both very strong.

Now, we can break down this observation into two others. Notice that, as we have been dealing with higher oxides, we have changed not only group numbers, but also the formula type of the oxide as we move across the series. Let us diminish the variables to one. Suppose we compare oxides of the same formula type; how do their acidic characters compare? Chlorine forms a gaseous oxide Cl_2O which forms the very weak hypochlorous acid in water (See Unit 11)

$$Cl_2O(g) + H_2O = 2HClO(aq)$$

but the oxide is obviously much more acidic than the very basic Na_2O. Again, SO_2 dissolves in water to give a fairly weak acid solution, but is obviously more acidic than SiO_2. These examples suggest that the acidity of oxides of the same formula type increases as we move across the period. It so happens that this is a sound generalization for all the short periods of the typical elements.

Now let us keep the group number constant and vary the formula type. If Cl_2O is weakly acidic, and Cl_2O_7 is the parent oxide of the very strong perchloric acid, then in this case, the higher oxide is more acidic than the lower one. Again, this is true of SO_2 and SO_3 (see above). Once more, this observation happens to be a sound generalization. Thus our initial generalization concerning the periodic variation of acid/basic properties can be broken down into two further generalizations:

1 The acid character of the oxides of the same formula type increases across a period of the typical elements.

2 For the normal oxides of the same element, higher oxides are more acidic than lower ones. (This applies universally throughout the Periodic Table.)

These trends are typical of those in which the Periodic Table abounds. Unfortunately, it is much less easy to explain them than to pick them out.

Now try SAQ 23 (p. 38).

---

* The solubility of the hydroxide or oxide in alkali is very important in the preparation of pure aluminium oxide for the extraction process. Bauxite is heated with an alkaline solution. Aluminium oxide dissolves in the alkali while impurities such as iron oxide do not. After filtration of the solution, the pure hydroxide can be reprecipitated and heated, when the anhydrous oxide is formed: $2Al(OH)_3 = Al_2O_3 + 3H_2O$.

19

6.2.4 Salts of sodium, magnesium and aluminium

Salts of sodium, magnesium and aluminium usually can be prepared by dissolving the metals in acids, or by neutralizing acids with the hydroxides, and then evaporating the solution. In Unit 6, Section 6.1.8, we mentioned that, in compounds at room temperature, sodium is confined to oxidation state $+1$, magnesium to oxidation state $+2$ and aluminium to oxidation state $+3$. In Table 4 is a list of acids classified by their basicity (the number of replaceable hydrogen atoms (see S100, Unit 9, Section 9.10)).

Table 4 Some acids that might form sodium, magnesium and aluminium salts

Acid	Basicity	Na	Mg	Al
HCl HNO_3 $HClO_4$ H_3PO_2	monobasic			
H_2CO_3 H_2SO_4 H_3PO_3	dibasic			
H_3PO_4	tribasic			

For each acid, write in Table 4 the formulae of the sodium, magnesium and aluminium salts which contain the greatest proportion of the metal.

Salts are formed by the replacement of hydrogen by sodium, and since the oxidation numbers of sodium and hydrogen are $+1$, one sodium replaces one hydrogen. The salt with the maximum proportion of sodium is the one in which all replaceable hydrogen atoms are replaced, e.g. $NaCl$, $NaNO_3$, $NaClO_4$, NaH_2PO_2, Na_2CO_3, Na_2SO_4, Na_2HPO_3, Na_3PO_4.

Likewise, as the oxidation numbers of magnesium and aluminium are $+2$ and $+3$ respectively, one magnesium replaces *two* hydrogens and one aluminium, *three* hydrogens.

The completed Table 4 is shown as Table 5.

Table 5 Formulae of sodium, magnesium and aluminium salts

Acid	Basicity	Na	Mg	Al
HCl HNO_3 $HClO_4$ H_3PO_2	monobasic	$NaCl$ $NaNO_3$ $NaClO_4$ NaH_2PO_2	$MgCl_2$ $Mg(NO_3)_2$ $Mg(ClO_4)_2$ $Mg(H_2PO_2)_2$	$AlCl_3$ $Al(NO_3)_3$ $Al(ClO_4)_3$ $Al(H_2PO_2)_3$
H_2CO_3 H_2SO_4 H_3PO_3	dibasic	Na_2CO_3 Na_2SO_4 Na_2HPO_3	$MgCO_3$ $MgSO_4$ $MgHPO_3$	$Al_2(CO_3)_3$ $Al_2(SO_4)_3$ $Al_2(HPO_3)_3$
H_3PO_4	tribasic	Na_3PO_4	$Mg_3(PO_4)_2$	$AlPO_4$

This exercise illustrates one important use of oxidation numbers: the knowledge that a metal exists in one particular oxidation number is a condensed expression of the formulae of many compounds. It might be argued that the formulae in Table 5 are equally well obtained by assuming that the compounds contain cations with inert gas structures (Na^+, Mg^{2+} and Al^{3+}). However, as we shall see, in this Unit and Unit 7, compounds such as $MgCl_2$ and $AlCl_3$ do not behave entirely in the manner we would expect of ionic compounds. Thus oxidation numbers express relationships between chemical formulae that transcend the boundaries of ionic-bonding models; consequently, they cannot in general be taken as an indication of the charge that an element carries.

When aqueous solutions of salts like those in Table 5 are evaporated, white solids are obtained. These often have the formulae given in the Table, but sometimes they contain water molecules. Thus sodium chloride, if it crystallizes from solution at 0 °C has the formula $NaCl.2H_2O$: every mole of sodium chloride contains two moles of water. Such substances are called *hydrated* salts or *hydrates*, to distinguish them from *anhydrous* salts like NaCl which contain no water. A hydrate can usually be converted into the anhydrous salt by heating it, when the water is driven off as steam.

hydrates

Some magnesium and aluminium compounds are insoluble in water, but sodium compounds, including sodium salts, are remarkable in that they all dissolve in water. Indeed, as a class, the alkali metals form very few insoluble salts.

6.2.5 The crystal structure of sodium nitrate*

The chief characteristics of those structures we call ionic are present in that of sodium chloride (shown as MgO in Figure 1). Each type of ion has a high coordination number** (in this case, six) to ions of opposite charge, and discrete molecules cannot be picked out. In NaCl the ions are monatomic, and it is natural to ask whether the characteristics of an ionic lattice are maintained when we introduce a complex anion like NO_3^-. This can be examined by using the example of sodium nitrate.

coordination number

The structure of sodium nitrate is related to that of sodium chloride, and is shown in Figure 5. Suppose the unit cell of NaCl is stood on one of its corners with a cube diagonal vertical. If the chlorides are replaced by planar NO_3 groups with the shape shown in Figure 6 in such a way that the NO_3 planes are horizontal (perpendicular to the vertical diagonal), something very close to the $NaNO_3$ structure results. The difference can be related to the fact that, unlike the chlorines in NaCl, the NO_3 groups have not got spherical symmetry. Consequently, the structure in Figure 5 is relatively extended in the horizontal plane and no longer has cubic symmetry. It retains only one of the four cubic threefold axes of symmetry***—the one along the vertical diagonal. Such a unit cell is described as rhombohedral.

axis of symmetry

The nearest neighbours of each sodium in the structure are six oxygens, one from each of the six surrounding NO_3 groups. The Na–O distance is about 250 pm, while the N–O distance in the NO_3 group is only 122 pm so the latter is easily picked out as a distinct unit. Figure 5 therefore illustrates several interesting

---

* Sodium nitrate is unique among chemical compounds; it was the cause of major war. You should already know from S100 that, as Chile saltpetre, it was a key source of fixed nitrogen for explosives and fertilizers during the nineteenth century before the invention of the Haber process.

Sodium nitrate is quite soluble in water, and it occurs naturally only in very arid regions like those of the Atacama desert in what is now Northern Chile. When the South American republics were created by the collapse of the Spanish empire during the period 1810–1825, some of their boundaries were not precisely defined. When the value of the nitrate deposits became obvious some years later, a dispute over their ownership arose between Chile on the one hand, and Peru and Bolivia on the other. War broke out in 1879. The territory of the belligerents was rugged, but their coastlines stretched nearly 4 000 miles from north to south. In these circumstances, it was inevitable that the war should be largely naval, and Chile possessed a strong maritime tradition dating back to successful operations against Spain in the wars of independence under Lord Cochrane, a distinguished British naval officer. By 1883 when the war ended, Chile had achieved a complete victory, and Peru and Bolivia were forced to surrender most of the nitrate deposits to her. These concessions left Bolivia an entirely landlocked nation, a handicap that has never ceased to hamper her subsequent economic development.

** The coordination number of an atom or ion is the number of nearest neighbours that it has in a particular chemical structure. As described in the text, the sodium ion in $NaNO_3$ has a coordination number of six with respect to the oxygens that constitute its nearest neighbours.

*** A cube has four diagonals. Any one of these diagonals is a threefold axis of symmetry. This means that if the cube is positioned with one of the diagonals vertical and viewed from above, then after rotation by 120°, the view is indistinguishable from what it was before rotation. The term *threefold* is used because there are three indistinguishable views in each complete rotation of 360°.

points. First, the way in which we can sometimes pick out groupings which we can call anions, in spite of the fact that such groupings contain several atoms. Secondly, the tendency of alkali metal cations to adopt high coordination numbers whether the anions are simple or complex, and thirdly, the fact that sometimes the symmetry of an ion exercises an observable effect on the symmetry of the unit cell that contains it.

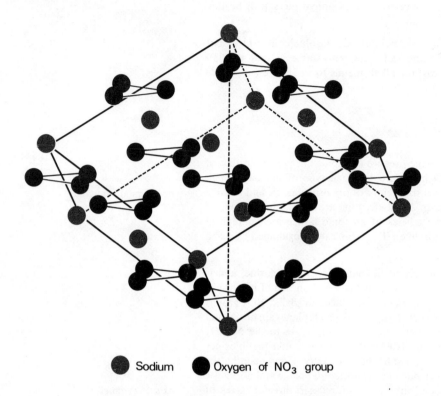

● Sodium   ● Oxygen of NO_3 group

Figure 5 Crystal structure of sodium nitrate.

The structure in Figure 5 is a fairly common one. It is also adopted by lithium nitrate, magnesium carbonate and the important 'calcite' form of calcium carbonate which is the chief constituent of limestone. In the last two cases the ions are doubly charged, the CO_3^{2-} ion, like the nitrate ion, having a planar shape.

6.2.6 Crystal structure of Epsom salts

If magnesium oxide, hydroxide or carbonate is dissolved in dilute sulphuric acid, and the solution is allowed to evaporate at room temperature, a solid hydrate of magnesium sulphate is left. This contains seven molecules of water of crystallization and is the famous 'Epsom salts', $MgSO_4 \cdot 7H_2O$.

The structure of this compound illustrates important points, not only about hydrates, but also about aqueous solutions.

We begin by taking a closer look at the structure of the water molecule itself. In S100, Unit 10, Section 10.5.2, the charge distribution in the molecule was examined. The oxygen atom has two lone pairs of electrons and also tends to attract electrons from the hydrogen. These two factors lead to a concentration of positive charge in the hydrogen regions of the molecule and a negative charge at the oxygen end.

Name one consequence of this.

Hydrogen bonds can be formed with other water molecules (S100, Unit 10, Section 10.5.2) or with negative oxygen atoms in other species. You may remember, too, the importance of hydrogen bonds in binding together the bases in the two different strands of DNA (S100, Unit 17, Section 17.2).

Now take a look at the structure of $MgSO_4 \cdot 7H_2O$ which is shown in projection in Figure 7. It seems very complex, so we will examine it piece by piece.

Figure 6 Planar skeleton of the nitrate ion. (*All three bonds are of equal length; their nature will be discussed in Unit 9*).

First, look at the red magnesium ion, then examine the key at the foot of the diagram. In the scale, $1\text{Å} = 100\text{pm} = 10^{-10}$ metres.

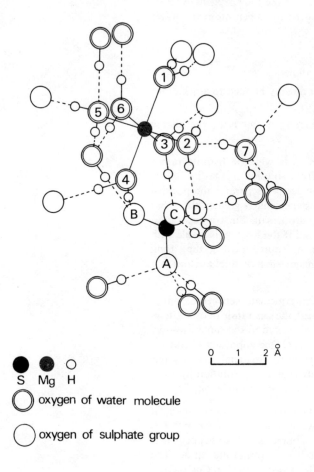

0   1   2 Å

● ● ○
S Mg H

◎ oxygen of water molecule

○ oxygen of sulphate group

Figure 7 Structure of Epsom salts.

What are the nearest atoms to this ion and how many are there?

They are the six oxygens marked 1–6 of six surrounding water molecules. The coordination of the oxygens around the magnesium is octahedral, because oxygens 2, 3, 5 and 6 form a square in a plane perpendicular to the line joining 1 and 4. The hydrogen atoms of the water molecules lie on the side of the molecule away from the magnesium.

Would you expect this?

Yes. If the magnesium is a positive ion, the negatively-charged oxygen region of the water molecule will tend to be closest to it, and the positive hydrogens further away. We can regard the magnesium ion as being attracted by the negative end of the water dipole (see Fig. 8).

$$\begin{array}{c} \delta+ \\ H \\ \diagdown \quad 2\delta- \\ \quad O\text{----}Mg^{2+} \\ \delta+ \diagup \\ H \end{array}$$

Figure 8 An interpretation of the magnesium–water bonding in $MgSO_4 \cdot 7H_2O$.

Do the water molecules 1–6 form bonds only to magnesium? Consider molecule 3 as a typical example.

No. Molecule 3 also forms two hydrogen-bonds through its two hydrogen atoms to two oxygens on different sulphate groups. All the other water molecules also form at least two hydrogen-bonds.

In what way does water molecule 7 differ from those marked 1–6?

It is not directly linked to a magnesium. It forms two hydrogen-bonds, one to a sulphate oxygen and one to the oxygen of water molecule 2. Two hydrogen-bonds are also formed to its oxygen by other water molecules that are attached to a magnesium which is not shown. Of the seven water molecules in $MgSO_4 \cdot 7H_2O$, six are linked to the magnesium, forming a grouping we can call $Mg(H_2O)_6^{2+}$. A more descriptive formula for the salt is $Mg(H_2O)_6\,SO_4 \cdot H_2O$.

What shape is the SO_4 grouping in the structure?

The black sulphur is surrounded by a tetrahedron of the four oxygens A, B, C and D. All sulphate oxygens are also linked to the hydrogen atoms of water molecules through hydrogen-bonding.

From this one structure we see that:

(a) water in hydrates can be linked directly to cations;

(b) water may also occupy sites where it is not directly bonded to any cations at all (but it is hydrogen-bonded to anions);

(c) the whole structure satisfies the fairly exacting demands of the charge distribution of the water molecule.

One important question that remains unanswered is why the hydrate rather than the anhydrous salt, $MgSO_4$, crystallized from solution in the first place. Figure 6 shows that the intervention of water molecules separates the positive magnesium and negative sulphate ions to a distance much greater than that which would presumably be found in the anhydrous salt. Such a separation absorbs energy which must somehow be recovered if the hydrate is stable with respect to the anhydrous salt plus water. The main source must be the strong electrostatic bonding which occurs between the magnesium or sulphate ions and their surrounding water molecules.

This suggests that a hydrate can be regarded as intermediate between an anhydrous salt and its solution. When a solution is formed, the separation of oppositely-charged ions is more extreme but the energy is recovered in the same way—by interaction between the ions and the water molecules. One important characteristic that distinguishes the typical metals from the transition metals is that water molecules surrounding their cations are not easily replaced in solution by some competing group such as an ammonia molecule, NH_3, or a cyanide ion, CN^-.

Such replacement often occurs much more readily in the case of transition metal ions, and with dramatic results, because, as we mentioned in Section 6.1.8, transition metal ions and compounds are often coloured, and it so happens that this colour is strongly dependent on the groups that surround the cation. The blue colour of aqueous solutions of copper salts is caused by the ion $Cu(H_2O)_4^{2+}$ in which the copper and the four oxygens of the water molecules lie in the same plane. When aqueous ammonia is added to such a solution, the much more intense colour of a new complex cation is produced:

$$Cu(H_2O)_4^{2+}(aq) + 4NH_3(aq) = Cu(NH_3)_4^{2+}(aq) + 4H_2O$$

In the Home Experiments, copper is contrasted in this respect with magnesium.

6.2.7 An unusual oxidation state of sodium

In this Unit, we have already spent a good deal of time considering the pattern of oxidation states as we move across a period of typical elements. Thus we know that if we burn sodium, magnesium and aluminium in chlorine, we obtain the compounds $NaCl$, $MgCl_2$ and $AlCl_3$.

It is very useful to examine this pattern from a quite different perspective: the simple progression in formulae is apparent not just because we obtain the compounds that we do, but because there are also many compounds that we do *not* obtain. For example, we do not observe the compounds Na_2Cl, $NaCl_2$ or $NaCl_3$, the compounds $MgCl$ or $MgCl_3$, or the compounds $AlCl$, $AlCl_2$ or $AlCl_4$.

In this Section we shall examine one of these compounds more closely, the compound $NaCl_2$, and we shall relate its non-existence to important thermodynamic properties of the sodium atom and its ions. Then in the next Section, we shall apply a similar treatment to $MgCl$ and $AlCl$. The treatment involves the Born-Haber cycle which you met for the first time in Unit 5, Section 5.8.

In Figure 9, you will find the Born-Haber cycle for a compound with the general formula MCl_2. This figure should be quite familiar if you did SAQ 18 in Unit 5.

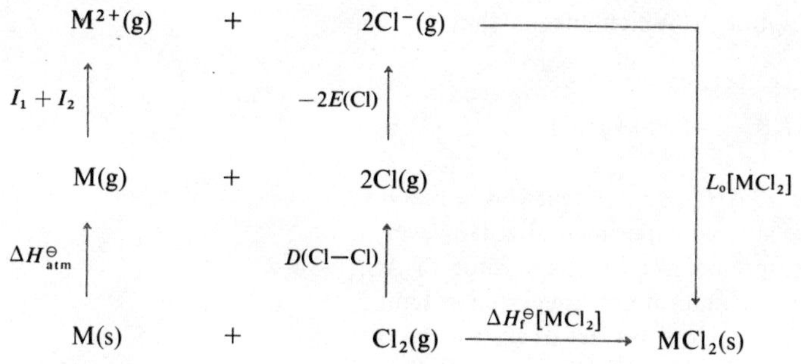

$$M^{2+}(g) \quad + \quad 2Cl^-(g) \xrightarrow{\hspace{4cm}}$$

$$I_1 + I_2 \uparrow \qquad\qquad -2E(Cl) \uparrow \qquad\qquad\qquad L_0[MCl_2] \downarrow$$

$$M(g) \quad + \quad 2Cl(g)$$

$$\Delta H^{\ominus}_{atm} \uparrow \qquad\qquad D(Cl-Cl) \uparrow$$

$$M(s) \quad + \quad Cl_2(g) \xrightarrow{\;\Delta H^{\ominus}_f[MCl_2]\;} MCl_2(s)$$

Figure 9 Born-Haber cycle for a compound MCl_2.

From the cycle, we can derive the following equation* by the First Law of Thermodynamics:

$$\Delta H^{\ominus}_f (MCl_2,s) = \Delta H^{\ominus}_{atm} + I_1 + I_2 + D(Cl-Cl) - 2E(Cl) + L_0$$

In Table 6, the experimentally measured quantities in this equation are given for $NaCl_2$ and $MgCl_2$.

Fill in the value of L_0 for $MgCl_2$.

The bottom number is the sum of those above it, so

$$L_0[MgCl_2] = -2\,497 \text{ KJ mol}^{-1}.$$

Table 6   Terms in the Born-Haber cycles for $MgCl_2$ and $NaCl_2$ ($kJ\,mol^{-1}$)

	Na	Mg
$\Delta H^{\ominus}_{atm}$	108	148
I_1	494	736
I_2	4 565	1 452
$D(Cl-Cl)$	244	244
$-2E(Cl)$	-724	-724
$L_0(MCl_2)$		
$\Delta H^{\ominus}_f (MCl_2,s)$		-641

At this stage we can introduce the assumption that both $MgCl_2$ and our hypothetical $NaCl_2$ contain ions of the type M^{2+} and Cl^-. Now sodium is nextdoor to magnesium in the Periodic Table and the ion Na^{2+} has the same charge, and only one less proton and one less electron than Mg^{2+}. Let us therefore assume that the two ions have similar sizes.

We attribute the decrease in energy when gaseous ions come together to form a solid lattice to the interaction between oppositely charged ions. This decrease will be larger the smaller the ions are, because then they pack more closely together. When gaseous ions come together in the process

$$M^{2+}(g) + 2Cl^-(g) \longrightarrow MCl_2(s)$$

to form solids $NaCl_2$ and $MgCl_2$ with similarly sized cations and the same type of ion arrangement, the internal energy or enthalpy changes should be much the same. Thus we might hope that $L_0[MgCl_2] = L_0[NaCl_2]$ to a good approximation.

Make this substitution in Table 6 and calculate $\Delta H^{\ominus}_f(NaCl_2,s)$.

* As in Unit 5, the terms I_1, I_2 E(Cl) and L_0 which refer to internal energy changes at 0 K have been used as enthalpy changes at 298 K. This introduces very little error.

You should get $\Delta H_f^{\ominus}(\text{NaCl}_2,\text{s}) = 2\,190\,\text{kJ mol}^{-1}$. Which means that for the reaction

$$\text{NaCl}_2(\text{s}) \longrightarrow \text{Na}(\text{s}) + \text{Cl}_2(\text{g}), \Delta H_m^{\ominus} = -2\,190\,\text{kJ mol}^{-1} \text{ at } 25\,°\text{C}.$$

We have now obtained an estimated value of $\Delta H_m^{\ominus}$ for this reaction, a reaction which involves a compound that cannot be studied experimentally. However, if we are interested in the stability of the compound, we require a value of $\Delta G_m^{\ominus}$ for the reaction. We could use the crude approximation first suggested in Unit 5, Section 5.7 and take $\Delta H_m^{\ominus}$ as a measure of $\Delta G_m^{\ominus}$. However, let us instead use the more refined approximation which was introduced in Section 5.7. Both the reactions

$$\text{NaCl}_2(\text{s}) = \text{Na}(\text{s}) + \text{Cl}_2(\text{g})$$

and

$$\text{MgCl}_2(\text{s}) = \text{Mg}(\text{s}) + \text{Cl}_2(\text{g})$$

are analogous in the sense used in Section 5.7, i.e. they both involve the decomposition of a solid to give a solid metal and *one mole of chlorine gas*. Thus the values of $\Delta S_m^{\ominus}$ should be similar.

Use the thermodynamic data in the *Data Book** to calculate $\Delta S_m^{\ominus}$ at 298 K for the reaction $\text{MgCl}_2(\text{s}) \longrightarrow \text{Mg}(\text{s}) + \text{Cl}_2(\text{g})$.

$$\Delta S_m^{\ominus} = 32\cdot7 + 223\cdot0 - 89\cdot6$$
$$= 166.1 \text{ J k}^{-1}\text{ mol}^{-1}$$

Now use this as an estimate for $\Delta S_m^{\ominus}$ for the sodium reaction and calculate $\Delta G_m^{\ominus}$ for this reaction at 25 °C.

$$\Delta G_m^{\ominus} = \Delta H_m^{\ominus} - T\Delta S_m^{\ominus}$$
$$= -2\,190 - 0\cdot298 \times 166\cdot1$$
$$= -2\,240 \text{ kJ mol}^{-1}$$

This suggests that a solid, NaCl_2, would be very unstable with respect to sodium and chlorine. As sodium and chlorine react readily to form NaCl, a more appropriate decomposition reaction is

$$\text{NaCl}_2(\text{s}) \longrightarrow \text{NaCl}(\text{s}) + \tfrac{1}{2}\text{Cl}_2$$

for which

$$\Delta G_m^{\ominus} = \Delta G_f^{\ominus}(\text{NaCl},\text{s}) - \Delta G_f^{\ominus}(\text{NaCl}_2,\text{s})$$

$$= -384 - 2\,240$$

$$= -2\,624 \text{ kJ mol}^{-1}$$

Again, our estimate suggests that the compound NaCl_2 is extremely unstable.

It is interesting to compare the two columns in Table 6 and to see which quantity tends to produce the very positive value of $\Delta G_f^{\ominus}(\text{NaCl}_2,\text{s})$. It is obviously the very high value of I_2 for sodium which corresponds to the change

$$\text{Na}^+(\text{g}) \longrightarrow \text{Na}^{2+}(\text{g}) + \text{e}^-(\text{g})$$

Here, the noble gas structure of the Na^+ ion loses an electron, so the large value of I_2 is a quantitative measure of the exceptional stability of this particular noble gas structure with respect to the loss of an electron. When the problem of the non-existence of sodium in oxidation states greater than one is examined in the way used here, the second ionization step in the cycle tends to produce a very positive value of $\Delta H_f^{\ominus}$. Consequently, the resulting compound is very unstable with respect to compounds in oxidation state one. This fact is obviously related to the tendency of sodium to occur in only one oxidation state. We begin to see, therefore, that the ionization potentials strongly influence the pattern of oxidation states for sodium, magnesium and aluminium.

---

* The Open University (1973) S24-/S25- *The Open University Chemistry Data Book*, The Open University Press.

6.2.8 An unusual oxidation state for magnesium and aluminium

A rather different case from the one considered in the previous Section is provided by the non-existence of solid compounds containing magnesium or aluminium in oxidation state one. We can again consider solid chlorides, in this instance MgCl and AlCl.

The Born-Haber cycle for a chloride, MCl, was considered in Unit 5, Section 5.8. It leads to the equation:

$$\Delta H_f^\ominus(MCl,s) = \Delta H_{atm}^\ominus + I_1 + \tfrac{1}{2}D(Cl-Cl) - E(Cl) + L_0[MCl]$$

In Table 7, the experimentally measured quantities in this equation are given for NaCl, MgCl and AlCl. As in the previous Section, we shall assume that the compounds are composed of ions, the three singly charged cations having the same size.

Estimate values of $\Delta H_f^\ominus$ for solid MgCl and AlCl.

$L_0[NaCl] = -773$ kJ mol^{-1}. Using this value for MgCl and AlCl,

$$\Delta H_f^\ominus(MgCl,s) = -129 \text{ kJ mol}^{-1}$$

and
$$\Delta H_f^\ominus(AlCl,s) = -110 \text{ kJ mol}^{-1}$$

From the *Data Book*, for the formation reaction

$$Na(s) + \tfrac{1}{2}Cl_2(g) \longrightarrow NaCl(s)$$

$$\Delta S_m^\ominus = 72.4 - 111.5 - 51.2$$

$$= -90.3 \text{ J K}^{-1} \text{ mol}^{-1}$$

We can use this value as an estimate for $\Delta S_f^\ominus(MgCl)$ and $\Delta S_f^\ominus(AlCl)$ and thus calculate values of $\Delta G_f^\ominus$ at 25°C.

Table 7   Terms in the Born-Haber cycles for NaCl, MgCl and AlCl (kJ mol^{-1})

	Na	Mg	Al
$\Delta H_{atm}^\ominus$	108	148	326
I_1	494	736	577
$\tfrac{1}{2}D(Cl-Cl)$	122	122	122
$-E(Cl)$	−362	−362	−362
$+L_0(MCl,s)$			
$\Delta H_f^\ominus(MCl,s)$	−411		

For MgCl,

$$\Delta G_f^\ominus = -129 + (0.298 \times 90.3) \text{ kJ mol}^{-1}$$

$$= -102 \text{ kJ mol}^{-1}$$

For AlCl,

$$\Delta G_f^\ominus = -110 + (0.298 \times 90.3) \text{ kJ mol}^{-1}$$

$$= -83 \text{ kJ mol}^{-1}$$

Do these negative values suggest that MgCl(s) and AlCl(s) are stable?

The question is incomplete: it does not ask what reaction the solids might be stable with respect to. The negative values of $\Delta G_f^\ominus$ suggest that the compounds are stable with respect to decomposition into their elements.

Can you suggest a reaction with respect to which MgCl might be unstable?

As Mg and $MgCl_2$ are well known, one possibility is

$$2MgCl(s) = Mg(s) + MgCl_2(s)$$

Using the *Data Book* and our estimate of $\Delta G_f^\ominus(MgCl,s)$

$$\Delta G_m^\ominus = -592 + (2 \times 102) \text{ kJ mol}^{-1}$$

$$= -388 \text{ kJ mol}^{-1}$$

likewise for the reaction

$$3AlCl(s) = 2Al(s) + AlCl_3(s), \quad \Delta G_m^\ominus = -380 \text{ kJ mol}^{-1}$$

In these two decomposition reactions, some metal atoms have their oxidation number increased while others *of the same type* have their oxidation numbers decreased. Such processes are called *disproportionation* reactions.

disproportionation

Calculations carried out in this Section suggest that, at normal temperatures, solid compounds containing magnesium or aluminium in oxidation state one should be thermodynamically *stable* with respect to their elements, but thermodynamically *unstable* with respect to disproportionation into the metal and a compound of a more usual oxidation state.

If aluminium and aluminium trichloride are heated to about $1\,000\,°C$, a *gaseous* compound AlCl is formed. Thus if we are prepared to use exceptional conditions, we can obtain aluminium, and indeed other elements, in exceptional oxidation states. It remains true, however, that under everyday conditions $+3$ is the normal oxidation state of aluminium in its compounds. In accordance with this generalization, if the AlCl formed at $1\,000\,°C$ is cooled to room temperature it disproportionates to the metal and trichloride before the gas can condense to form solid AlCl. This behaviour is in accord with our predicted value of $\Delta G_m^\ominus$ for the disproportionation at $25\,°C$.

6.2.9 Chlorides

If solutions of the chlorides of sodium, magnesium and aluminium are prepared by neutralization reactions, and then allowed to evaporate in a warm room, only the sodium compound is obtained as the anhydrous salt. From the magnesium solution, a hexahydrate $MgCl_2.6H_2O$ is obtained. Moreover, the anhydrous chloride cannot be obtained by the usual method of heating the hydrate, because when heated, it reacts with water to form oxygen compounds, e.g.

$$MgCl_2 + H_2O = MgO + 2HCl$$

The neutral solution of aluminium trichloride does not even deposit a pure *hydrated* chloride. The crystalline product contains a substantial amount of the hydroxide. For these reasons, the anhydrous chlorides of magnesium and aluminium are best prepared by the reaction of the metals with dry chlorine or dry hydrogen chloride. You saw such a preparation in the TV programme for Unit 3 when we made $AlCl_3$.

After this preparation, you were shown a property of the trichloride which is related to the behaviour of the solution described above.

What was this property?

The trichloride reacted with moist air at room temperature giving the hydroxide and hydrogen chloride gas:

$$AlCl_3(s) + 3H_2O = Al(OH)_3(s) + 3HCl(g) \tag{9}$$

Such a reaction is consistent with the fact that when we try to obtain $AlCl_3$ from its aqueous solution, our solid product is contaminated with aluminium hydroxide.

Notice that equation 9 is the reverse form of a neutralization. A salt reacts with water to give an acid gas, HCl, and aluminium hydroxide, which can act as a base. Such a process is called the *hydrolysis* of a salt. Although the hydroxide dissolves in a *solution* of hydrogen chloride to give a dilute *solution* of the trichloride, the *solid* trichloride will react with water to give the hydroxide and

hydrolysis

gaseous hydrogen chloride. This shows that when the states of reactants and products are changed, the position of equilibrium may also change quite markedly.

As we move from sodium to magnesium to aluminium, the stability of the trichlorides with respect to hydrolysis seems to decrease. The composition of salt is unchanged by melting in moist air, anhydrous magnesium chloride picks up water from the air and loses hydrogen chloride when heated, and (as we have seen) aluminium trichloride fumes in moist air at room temperature. The bromides and iodides of sodium, magnesium and aluminium show a similar tendency.

Another interesting characteristic of aluminium trichloride was apparent from the TV programme for Unit 3. Sodium and magnesium chlorides have the involatility that we associate with 'ionic' compounds, but it was clear during its preparation that aluminium trichloride could be easily vaporized, and in fact, this occurs at about 180°C. The vapour has some unusual properties. It can be shown that it consists of discrete gaseous molecules with the formula Al_2Cl_6. The molecular mass can be established by carefully weighing a known volume of the gas at a particular temperature and pressure and, in the Home Experiment for Unit 11, you will apply this method to a volatile liquid. Here, we shall concentrate on the results for vapourized aluminium trichloride. The formula Al_2Cl_6 shows that in the vapour, two molecules of $AlCl_3$ are linked together in some way.

The volatility of $AlCl_3$, which involves the formation of what can only be a covalent molecule of formula Al_2Cl_6, suggests that we should think twice before describing solid $AlCl_3$ as 'ionic'. You will recall, too, from Unit 3, that fused $AlCl_3$ is a non-conductor. In the next Unit we shall examine the structure of solid $AlCl_3$, and see if *this* has the characteristics that we would expect of a solid composed of ions.

6.2.10 The thermal stability of the salts of oxy-anions

Consider the effect of heating sodium and magnesium carbonates. At about 400°C, magnesium carbonate loses carbon dioxide as in the reaction

$$MgCO_3 = MgO + CO_2$$

but Na_2CO_3 can be melted at over 800 °C without decomposition.

If one tries to prepare *aluminium* carbonate by adding sodium carbonate to a solution of an aluminium salt, the precipitate turns out to be aluminium hydroxide. Moreover, all other attempts to prepare the compound, for example, by exposing Al_2O_3 to carbon dioxide at, or just above room temperature, also fail. These experiments make it likely that pure aluminium carbonate does not exist, because at normal temperatures it is extremely unstable with respect to the reaction

$$Al_2(CO_3)_3 \longrightarrow Al_2O_3 + 3CO_2$$

Thus the carbonates become less stable with respect to the oxides and CO_2 as one moves from sodium to aluminium.

Can you recall a similar stability progression from earlier in the Unit?

The behaviour of the peroxides (Section 6.2.2) is consistent with such a progression.

Again, if the sulphates are prepared by crystallization from aqueous solutions, they form hydrates (compare with Section 6.2.6), but when these are heated, water is driven off, and the anhydrous salts, Na_2SO_4, $MgSO_4$ and $Al_2(SO_4)_3$ are left. If heating is intensified, the aluminium sulphate is the first to decompose, the sulphate losing oxides of sulphur at about 600°C in reactions of the type

$$Al_2(SO_4)_3 = Al_2O_3 + 3SO_3(g)$$

At about 1 000 °C, the magnesium compound behaves similarly, leaving sodium sulphate undecomposed.

Both the decreasing stability of the salts of oxy-anions with respect to a volatile product and the oxide, as one moves from sodium to aluminium, and the decreasing stability of halides with respect to hydrolysis, fit in well with one theme from the TV programme for Unit 3: the prominence of oxide formation in the chemistry of aluminium and its compounds.

6.3 The alkali metals and alkaline earth metals as groups

The examination of the chemistry of sodium, magnesium and aluminium has so far channelled our interest into the changes that occur as we move across a period of the typical elements. This emphasis could be criticized on the grounds that the chemical variations within a Group are very important. However, in the case of Groups I and II, we have a reasonable defence: within these two Groups, the elements resemble one another much more closely than those of Groups III to VII. This means that by discussing the chemistry of sodium and magnesium, we come reasonably close to a general description of the elements of Groups I and II.

Let us elaborate on this point a little. First of all, notice that the elements of Groups I and II are all metals. What is more, the metals are all very 'electropositive' in the sense in which the word was used in Unit 5, i.e. they occur at the top of the various series of metals discussed there, and they combine readily with acids, oxygen and the halogens in strongly exothermic reactions. Finally, the chemistry of their compounds is confined to that of only one oxidation state. In the case of the alkali metals this is $+1$, and in the case of the alkaline earths, $+2$. Sweeping statements of this kind cannot be made for the elements of later Groups. For example, in the case of Group VI, oxygen is a non-metallic colourless gas whose highest fluoride is OF_2, while polonium is a metal which can form a hexafluoride PoF_6.

Nevertheless, less striking but clearly discernible changes do occur when we move down the Groups of the alkali or alkaline earth metals. One of these is the change in the thermal stabilities of compounds.

6.3.1 The thermal stability of alkali metal and alkaline earth metal compounds

In Section 6.2.10, we discussed the thermal stability of some sodium, magnesium and aluminium compounds in terms of their decomposition temperatures. For example, we pointed out that magnesium carbonate decomposes at about 400 °C into magnesium oxide and carbon dioxide. This decomposition is typical of all the alkaline earth metal carbonates, but for each metal the decomposition temperature of the carbonate is different. For example, in the case of barium, it is about 1400 °C, much higher than for magnesium. We shall now try to relate the decomposition temperatures to thermodynamic quantities.

In Table 8 you will find the values of $\Delta G_m^\ominus$, $\Delta H_m^\ominus$ and $\Delta S_m^\ominus$ for the decompositions of four alkaline metal carbonates at 298.15 K. As you can verify, for these figures

$$\Delta G_m^\ominus = \Delta H_m^\ominus - T\Delta S_m^\ominus \tag{10}$$

Notice first that all the values of $\Delta G_m^\ominus$ are positive – at 298.15 K all the carbonates are thermodynamically stable with respect to the decomposition reactions. Thus when the compounds are heated, at some temperature $\Delta G_m^\ominus$ must become negative. How do the $\Delta H_m^\ominus$ and the $T\Delta S_m^\ominus$ terms conspire to achieve this?

You can see that the four values of $\Delta S_m^\ominus$ are very similar and positive, a situation that we might expect because the reactions are analogous in the sense in which the word was used in Unit 5, Section 5.7: they all involve the decomposition of a carbonate into an oxide, and the same number of moles of the same gas, carbon dioxide, are produced in each case. Because a gas is produced, the values of $\Delta S_m^\ominus$ are positive.

Now suppose we heat samples of calcium and barium carbonates in closed evacuated containers. Consider the terms on the right-hand side of equation 10.

What happens to the values of $\Delta H_m^{\ominus}$ as the temperature is increased?

As we pointed out in S100, Unit 12, values of $\Delta H_m^{\ominus}$ for a given reaction change very little with temperature, so we can use the value of $\Delta H_m^{\ominus}$ at 298.15 K for $\Delta H_m^{\ominus}$ at any temperature, T, and write

$$\Delta G_T^{\ominus} \simeq \Delta H_{298}^{\ominus} - T\Delta S_T^{\ominus} \qquad (11)$$

where the subscripts now denote temperature.

Table 8   Thermodynamic properties of the decomposition reactions of alkaline earth carbonates at 25°C

Reaction	$\Delta G_m^{\ominus}/$ kJ mol^{-1}	$\Delta H_m^{\ominus}/$ kJ mol^{-1}	$\Delta S_m^{\ominus}/$ J K^{-1} mol^{-1}
$MgCO_3 = MgO + CO_2$	65·5	117·7	175·5
$CaCO_3 = CaO + CO_2$	130·6	178·5	160·4
$SrCO_3 = SrO + CO_2$	186·4	237·4	170·9
$BaCO_3 = BaO + CO_2$	215·6	266·7	171·8

We also know that at any temperature T, the values of $\Delta S_T^{\ominus}$ for the calcium and barium carbonate decomposition should be nearly the same, that is, the values of $T\Delta S_T^{\ominus}$ should be nearly the same.

Now for both reactions, $\Delta S_m^{\ominus}$ is positive at 298 K, and indeed remains positive at all accessible temperatures. Thus, if we apply equation 11 to calcium carbonate, as T increases, there will come a point when $T\Delta S_T^{\ominus}$ is equal to $\Delta H_{298}^{\ominus}$ and $\Delta G_T^{\ominus}$ will be zero. When this is so, then *if equilibrium exists between the carbonate, the oxide and carbon dioxide gas*, appreciable decomposition will have taken place, and the temperature at which $\Delta G_T^{\ominus} = 0$ can be conveniently chosen to define the decomposition temperature of calcium carbonate. Above this temperature $\Delta G_T^{\ominus}$ becomes negative and decomposition is even more favourable.

At the same time, as $\Delta H_{298}^{\ominus}$ is larger for barium carbonate, the value of $T\Delta S_T^{\ominus}$ at the decomposition temperature of calcium carbonate will not be enough, when substituted into equation 11, to make $\Delta G_T^{\ominus}$ for $BaCO_3$ zero. In fact $\Delta G_T^{\ominus}$ for barium carbonate will still be positive, and a higher temperature will be required to bring it down to zero. We conclude, therefore, that for analogous decompositions like those in Table 8, with positive values of $\Delta G_m^{\ominus}$ and $\Delta S_m^{\ominus}$ at 298.15 K, the equilibrium decomposition temperature is likely to be the greater, the more positive $\Delta H_m^{\ominus}$ at 298.15 K.

To reinforce this point try SAQ 28 (on p. 39).

Because values of $\Delta H_{298}^{\ominus}$ become more positive as we move from magnesium to barium, the equilibrium decomposition temperatures of the carbonates should increase in this order too. This is so: the temperatures at which the equilibrium pressure of carbon dioxide above the carbonates reaches one atmosphere are 400 °C for magnesium, 900 °C for calcium, 1 280 °C for strontium and 1 360 °C for barium carbonate. This is related to the fact that when heated in the open atmosphere, the temperatures at which the onset of decomposition becomes perceptible increase in the sequence $MgCO_3$, $CaCO_3$, $SrCO_3$ and $BaCO_3$: the thermal stability of the carbonates increases down the group.

A similar example is provided by the alkaline earth sulphates which undergo decomposition reactions of the type

$$MSO_4(s) = MO(s) + SO_3(g)$$

where M is an alkaline earth metal.

Magnesium sulphate decomposes at about 1 000 °C, for example, and barium sulphate at about 1 600 °C. Finally, what is true for the alkaline earth carbonates is also true for those of the alkali metals. Table 9 shows the values of $\Delta H_{298}^{\ominus}$ for the decompositions

$$M_2CO_3(s) = M_2O(s) + CO_2(g)$$

where M is an alkali metal.

31

These values increase down the group. One result of this is that the lithium compound is the only alkali metal carbonate that decomposes at temperatures attainable in the flame of a Bunsen burner.

$$Li_2CO_3(s) = Li_2O(s) + CO_2(g)$$

It seems, therefore, that the thermodynamic stability of the salts of oxy-anions of the alkali and alkaline earth metals (with respect to the metallic oxides and a volatile product) increases down the group.

Table 9   Standard enthalpy changes for the reaction
$M_2CO_3(s) = M_2O(s) + CO_2(g)$ at 298.15 K

	Li	Na	K	Rb	Cs
$\Delta H_m^{\ominus}/kJ\ mol^{-1}$	225·9	321·3	390·8	404·2	407·5

It turns out, however, that this generalization is only part of a broader one that we can make. If we write the decompositions in an ionic form, we notice that the anions in the reactants contain more atoms than the anions in the products which are usually monatomic. Compare, for example, CO_3^{2-} with O^{2-} and SO_4^{2-} with O^{2-}; it is clear then that the product anions are likely to occupy less space in a crystal than the anions in the reactant. In general, it turns out that in decompositions where a solid alkali metal or alkaline earth salt decomposes to a solid compound *containing a smaller anion*, the stability of compounds with respect to the decomposition increases down the group.

In Unit 7 we shall explore the reasons for this trend by using simple models of ionic bonding, but in this Unit we shall only note factual examples.

We can now see that the formation of sodium peroxide when the metal burns in air (Section 6.1.2) is related to the trend because lithium gives the normal oxide, Li_2O, and not Li_2O_2 which is presumably unstable to loss of oxygen at the temperature of the combustion. On the other hand, potassium, rubidium and caesium take on even more oxygen than sodium; when they burn in air, orange solids with the formula MO_2 are formed. These contain the *superoxide* ion, O_2^-.

Different trends in stability are sometimes observed, but these occur with a different type of reaction. When warmed with hydrogen, the alkali metals form hydrides whose thermal stability with respect to the reaction

$$MH(s) = M(s) + \tfrac{1}{2}H_2(g)$$

decreases down the group. For example, lithium hydride melts at 700 °C without decomposing, but the other hydrides give off hydrogen below 500 °C.

Thus the alkali metal hydrides exhibit a reverse trend in stability to those that we have observed up until now. But then, the decomposition reaction is of a different nature; a salt decomposes not to form another salt with a smaller anion, but to form the metal itself.

6.4   Summary

In this Unit, we first examined the periodic variation in the formulae of certain types of compounds and, especially, the periodic variation in the formulae of the highest normal oxides which was of great importance in first establishing the periodic law. We discussed how the concept of valency, i.e. of a characteristic number of links that a particular atom could form with other atoms, arose out of relationships between the formulae of different classes of compounds and then we showed how the concept was weakened by structural information such as the atomic arrangement in sodium chloride.

The use of oxidation numbers retained some of the classifying powers of valency and, at the same time, improved the precision of the language used to discuss oxidation-reduction equations. The pattern of oxidation states for a typical element is usually different from that for a transition element.

In the second half of the Unit we looked at some of the detailed chemistry of sodium, magnesium and aluminium. We did this, not just because the three metals are very important elements, but because they offer particular examples and instances of important *general* principles and tendencies. This process was then continued in Section 6.3 and your attention is directed in particular to:

1   The importance of electrolytic methods in the preparation of the most 'electropositive' elements (Section 6.2.1).

2   The distinction between kinetic stability (stability caused by a slow rate of reaction) and thermodynamic stability (e.g. the reaction of magnesium and aluminium with water in Section 6.2.1).

3   The importance of associating the non-existence of a compound with a particular decomposition reaction (e.g. $Al_2(O_2)_3$, $NaCl_2$, $MgCl$, $AlCl(s)$, $Al_2(CO_3)_3$. See also the compounds in this Unit's TV programme.)

4   General comments on the solubility of hydroxides in water (Section 6.2.3).

5   Variations in the acidic and basic properties of oxides across a row of the typical elements and also with the oxidation state of each element (Section 6.2.3).

6   The prediction of formulae of compounds using generalizations concerning oxidation states (Section 6.2.4).

7   The way in which the characteristics of an ionic lattice can be preserved but modified when anions are complex (Section 6.2.5).

8   The positions occupied by water molecules in the structure of solid hydrates (Section 6.2.6).

9   The use of arguments concerning the interaction energies of ions in solids to calculate the enthalpies of decomposition of unknown compounds (e.g. $NaCl_2$, $AlCl$, $MgCl$).

10   The decreasing stability of halides with respect to hydrolysis, and of salts of oxy-anions with respect to oxide formation, as one moves from Group I to Group III (Sections 6.2.9 and 6.2.10).

11   The correlation between $\Delta H_{298}^{\ominus}$ and the decomposition temperature for analogous decomposition reactions at equilibrium (Section 6.3.1).

12   The increase in the thermal stability of salts down Groups Ia and IIa when the solid product contains a smaller anion (Section 6.3.1).

In this Unit, we found ourselves alluding, at crucial moments, to the sizes of ions and the energies of interaction between ions. We were forced to make purely qualitative comparisons, but in Unit 7, we shall begin to discuss attempts to put these concepts on a quantitative basis. At the same time, we shall be forced to examine the packing of ions in crystals and to see whether the structural evidence supports the description of some of the compounds in this Unit as 'ionic'. This leads us to a more careful examination of those bonding theories that are familiar to you.

Appendix 1 (White)

Chemical formulae and stoichiometry

This Appendix is partly a revision and partly an extension of the material in S100, Unit 6, Appendix 2. If you feel confident about the material, try SAQs 10, 11 and 12 which are the most broadly-based.

Except for SAQ 12, the problems have been made up so that the arithmetic is very simple—you should not need log tables, or a slide rule.

Chemical formulae

Chemical formulae can provide us with several different pieces of information.

1   A chemical formula reveals the relative numbers of atoms present in a compound. Thus the formula SiO_2 tells us that in silica, every atom of silicon is combined with two atoms of oxygen. Notice that the formulae Si_2O_4 or Si_5O_{10} do this equally well, but unless there is a good reason (see (4)) the numbers in a formula are reduced by dividing throughout by the highest common factor. This is two in the case of Si_2O_4 and five in the case of Si_5O_{10}. The formula obtained by this process is called the *empirical formula* or sometimes the 'simplest formula'.

>   SAQ 1 (Objective 1)   What are the empirical formulae for the formulae N_4S_4, $C_{18}H_{38}$, Al_2Cl_6, NaCl, C_6H_6 and $B_{20}H_{16}$?

>   The two carbon compounds may lead you to the good reason mentioned in 1 above.

2   The chemical formula of a compound reveals the numbers of moles of the different elements that one mole of a compound contains. Given the existence of the Avogadro number, the number of particles in one mole, this follows from 1 above. Thus, if in silica one atom of silicon is combined with two atoms of oxygen, then $1 \times 6.02 \times 10^{23}$ atoms of silicon must be combined with $2 \times 6.02 \times 10^{23}$ atoms of oxygen. That is, one mole of silicon atoms is combined with two moles of oxygen atoms in one mole of SiO_2.

>   SAQ 2 (Objective 2)   How many moles of oxygen atoms are there in 0·35 moles of $Al_2(SO_4)_3$?

>   SAQ 3 (Objective 2)   How many moles of hydrazine, N_2H_4, contain three moles of nitrogen atoms?

3   If you have a table of relative atomic masses at your elbow then you can use the formula of a compound to calculate the masses of the different elements in a known mass of the compound. For the problems that follow, relative atomic masses are given at the end of each question.

>   SAQ 4 (Objective 2)   Calculate the masses of chromium and fluorine in 4·50 g of $CrF_2(Cr = 52·0, F = 19·0)$.

A more common problem is to calculate the *empirical* formula of a compound from the relative masses of the elements it contains.

>   SAQ 5 (Objective 3)   Three gases are analysed and are found to have the following composition:

>   gas 1: 12 g of carbon to each 1 g of hydrogen
>   gas 2: 6 g of carbon to each 1 g of hydrogen
>   gas 3: 4 g of carbon to each 1 g of hydrogen.

>   What are their empirical formulae? $(C = 12, H = 1)$

>   Further examples of this type of calculation form part of later questions.

4   In certain cases, a chemical formula expresses more than just the *relative* numbers of atoms or masses of elements that a compound contains: it carries a structural implication. In many compounds that we call 'covalent' for example, we have evidence that discrete molecules exist. Thus CO_2 is regarded as the correct formula for carbon dioxide because a molecule of carbon dioxide contains one atom of carbon and two of oxygen. C_2O_4 would not be consistent with this. When a formula carries this implication it is called the *molecular* formula

of the compound. Giant lattices like salt and silica have no molecular formulae and Na_3Cl_3 or Si_2O_4, for instance, carry as much information as NaCl or SiO_2. The latter are chosen only because they are simpler.

> SAQ 6 (Objective 6) Suggest two possible molecular formulae for the compound with empirical formula CH in SAQ 5.

Chemical equations

Chemical formulae give us the relative numbers of atoms and relative masses of the elements which are combined together in compounds. Chemical equations give us the changes in these quantities when the compounds react together.

In establishing the equation for a chemical reaction, the first step is to establish by experiments the nature of the reactants and products. This means establishing their chemical formulae. For example, when hydrogen and chlorine molecules react together, it is found that hydrogen chloride, HCl, is formed:

$$H_2 + Cl_2 = HCl$$

There is an important omission from this equation. In *chemical* reactions, repeated experiment suggests that within the limits of accuracy of our most precise instruments:

(a) the total mass of reactants and products is unchanged in a chemical reaction;

(b) the nuclei with their number of protons characteristic of a particular element, and the electrons, retain their individual identity. Only their state of combination with each other changes.

This means that chemical equations must be *balanced*: the same number of atoms of each type must appear on both sides of the equation, and the charges on each side must add up to the same number. In the examples above, when one hydrogen molecule, H_2, and one chlorine molecule, Cl_2, are consumed, *two* molecules of hydrogen chloride, HCl, must be produced. Thus the balanced reaction is

$$H_2 + Cl_2 = 2HCl$$

> SAQ 7 (Objective 4) The following equations are not balanced. They give only the formulae of the reactants and products. Balance them.
>
> (a) $H_2S + SO_2 = S + H_2O$
> (b) $N_2O_4 + N_2H_4 = N_2 + H_2O$
> (c) $KClO_3 = KCl + KClO_4$
> (d) $C_5H_{12} + O_2 = CO_2 + H_2O$
> (e) $H_3PO_3 = H_3PO_4 + PH_3$
> (f) $Al_4C_3 + H_2O = Al(OH)_3 + CH_4$

Combining weights and yields in chemical reactions

We have already mentioned that a balanced chemical equation tells us the relative changes in the number of atoms or the number of moles of reactants and products when the reaction occurs.

> In the equation $4HCl + O_2 = 2H_2O + 2Cl_2$, how many moles of oxygen molecules (O_2) are required to convert 0·28 moles of HCl to chlorine and water?

0·07 moles. Four moles of HCl react with one of O_2, so 0·28 moles react with $\frac{1}{4} \times 0.28 = 0.07$ moles O_2.

> SAQ 8 (Objective 5) XeF_6 can be made by reacting xenon with fluorine. How many moles of XeF_6 can be made from 0·0393 g of xenon and 0·0456 g of fluorine? (Xe = 131, F = 19·0).

> SAQ 9 (Objective 5) Hydrazine, N_2H_4, reacts with hydrogen peroxide, H_2O_2 as follows:
>
> $$N_2H_4 + 2H_2O_2 = N_2 + 4H_2O$$
>
> (a) When 0·128 g of hydrazine is consumed in this reaction, what mass of water is produced?
>
> (b) What mass of hydrogen peroxide will be consumed at the same time? (N = 14·0, H = 1·0, O = 16·0).

35

The following problems combine several types of calculation, and are typical of the kinds of problem that might be encountered in establishing the formulae of unknown compounds.

SAQ 10 (Objectives 2, 3 and 5) A chloride of phosphorus reacts with water. 0·0834 g of the chloride is dissolved in dilute acid and silver nitrate solution is added. A white precipitate of silver chloride, AgCl, is formed. This is filtered off, washed and dried. It weighs 0·2868 g. It is found that the filtered liquid does not contain a detectable amount of chlorine in any form. What is the empirical formula of the chloride of phosphorus? (Ag = 107·9, Cl = 35·5, P = 31·0).

SAQ 11 (Objectives 2, 3 and 5) 0·4636 g of an oxide of silver is dissolved in dilute nitric acid. When dilute hydrochloric acid is added, all the silver is precipitated as AgCl. After filtering, washing and drying, the precipitate weighs 0·5736 g. What is the empirical formula of the oxide? (Ag = 107·9, Cl = 35·5, O = 16.)

SAQ 12 (Objectives 2, 3 and 5) 3·73 g of a chloride of platinum is heated. Chlorine gas is driven off and 2·17 g of platinum metal is left. What is the empirical formula of the chloride?

You will need log tables or a slide rule this time. The problem is realistic in that as in an experimental situation, you will not quite get a whole number for the ratio of chlorine atoms to platinum atoms. Use the relative atomic masses Pt = 195·1, Cl = 35·45.

Self-assessment questions

SAQs 1–12 covering Objectives 1–5, are to be found in Appendix 1.

SAQ 13 (Objective 6)   The atomic number of neptunium, Np, is one greater than that of uranium. What would you expect the formula of the highest normal oxide of this element to be?

SAQ 14 (Objective 6)   An element M is known to form the following compounds: oxides with the formulae M_2O_3, MO_2, M_2O_5, MO_3 and M_2O_7; fluorides with the formulae MF_4, MF_5, MF_6 and MF_7. Its atomic weight lies between that of barium (137) and lead (207). Try to identify the element.

SAQ 15 (Objective 7)   Nitrogen forms molecules with the following molecular formulae: NH_3, N_2O_3, NF_3, N_2F_2, N_2H_4, N_3H_5, NOF, N_2, NH_3O.

Can you represent all these compounds by diagrams in which nitrogen, hydrogen, oxygen and fluorine each have a fixed valency? (Do not worry about the shapes of the molecules).

SAQ 16 (Objective 8)   Assign oxidation numbers to each of the elements in the following compounds and ions:

(a) I_2
(b) U^{3+}
(c) $BaCl_2$
(d) LiH
(e) H_2SO_3

(f) H_6TeO_6
(g) $Ga_2(SO_4)_3$
(h) HOI
(i) IO_6^{5-}
(j) VO_2^{+}

SAQ 17 (Objective 10)

(a)   Two elements L and M form aqueous cations:

$$L \text{ forms } L^+ \text{ (aq) and } L^{3+} \text{ (aq),}$$

$$M \text{ forms } M^{2+} \text{ (aq) and } M^{3+} \text{ (aq)}$$

One element is a typical element and the other is a transition metal. Suggest which is which.

(b)   Two elements, L and M form the following oxides and oxy-anions:

L forms $H_2LO_2^-$, $H_2L_2O_6^{2-}$, HLO_3^{2-}, LO_3^-, LO_4^{3-}, L_2O_3 and L_2O_5;

M forms MO, M_2O_3, MO_2, MO_4^{3-}, MO_4^{2-} and MO_4^-.

One element is a typical element and the other is a transition metal. Suggest which is which.

SAQ 18 (Objective 9)   Produce balanced equations for the following changes in an acid aqueous solution:
(a) $Cr_2O_7^{2-} + Fe^{2+} = Cr^{3+} + Fe^{3+}$
(b) $IO_3^- + I^- = I_2$
(c) $Cu + NO_3^- = Cu^{2+} + NO$
(d) $Al + NO_3^- = Al^{3+} + NH_4^+$
(e) $MnO_4^- + H_2O_2 = Mn^{2+} + O_2$

SAQ 19 (Objective 11)   Classify the following reactions as either (a) oxidation of an oxidation state of manganese (b) reduction of an oxidation state of manganese (c) disproportionation of an oxidation state of manganese (d) neither (a), (b) or (c).

(i) $3MnO_4^{2-} + 4H^+ = 2MnO_4^- + MnO_2 + 2H_2O$
(ii) $MnO_4^{2-} + H_2O_2 = MnO_2 + O_2 + 2OH^-$
(iii) $2Mn^{3+} + 2H_2O = Mn^{2+} + MnO_2 + 4H^+$
(iv) $Mn(OH)_2 + 2H^+ = Mn^{2+} + 2H_2O$
(v) $2MnO_4^{2-} + S_2O_8^{2-} = 2MnO_4^- + 2SO_4^{2-}$

SAQ 20 (Objectives 3 and 13)   One of the alkaline earth metal hydrides contains less than 2 per cent hydrogen by weight and is not radioactive. Which one is it? (Relative atomic masses are Be $=$ 9, Mg $=$ 24, Ca $=$ 40, Sr $=$ 88, Ba $=$ 137, Ra $=$ 226 and H $=$ 1.)

SAQ 21 (Objectives 12 and 13)   Two of the six alkaline earth metals in the previous questions do not form anhydrous peroxides. Which ones would you suggest?

SAQ 22 (Objective 13)   Use the thermodynamic data given in Table 7 together with the first ionization energy of the inert gas neon ($2\,079$ kJ mol^{-1}) to estimate the standard enthalpy of formation of an ionic solid with the formula NeCl. Comment on the stability of the compound with respect to a possible decomposition reaction. Does your answer square with the reluctance of the noble gases to form compounds?

SAQ 23 (Objective 13)   Equimolar amounts of the following pairs of oxides are added to water. In each case, state which solution you would expect to have the lowest pH.

(a) Cr_2O_3 and CrO_3
(b) GeO_2 and SeO_2
(c) Rb_2O and In_2O_3
(d) Sb_2O_3 and I_2O_5

SAQ 24 (Objective 13)   Magnesium and hydrogen do not react at room temperature. Is this because magnesium hydride is thermodynamically unstable with respect to magnesium and hydrogen at room temperature? At $298\cdot15$ K, $\Delta G_f^{\ominus}$ (MgH_2,s) $= -34\cdot6$ kJ mol^{-1}.

SAQ 25 (Objective 13)   On burning in air, potassium forms a compound KO_2, potassium superoxide. This may be regarded as containing the ions K^+ and O_2^-. The unit cell may be obtained by replacing the Na^+ ions in NaCl with K^+ ions and the Cl^- ions in NaCl by O_2^- ions so that the O–O axes are parallel and lie along a cell edge (see Fig. 10).

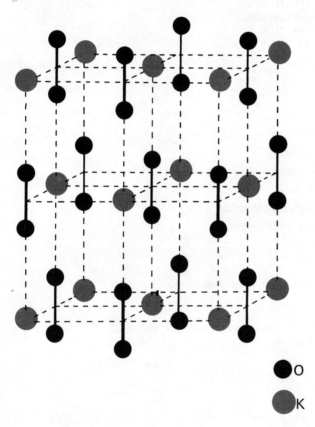

O

K

Would you expect the unit cell to remain cubic after such a substitution? If not, how would you expect it to change?

SAQ 26 (Objective 13)   You wish to make a sample of barium metal from BaO, a solid with very high melting point. Would you recommend reducing the oxide with heated carbon? If not, suggest an alternative procedure.

SAQ 27 (Objective 13)   Until fairly recently, compounds containing the ClF_4^- ion were, unknown. A possible way of preparing them is to react an alkali metal fluoride with gaseous ClF_3, using a temperature of about $100\,^{\circ}$C to speed up the reaction:

$$MF\ (s) + ClF_3(g) = MClF_4\ (s)$$

Which alkali metal fluoride would you choose?

SAQ 28 (Objective 13)

At 927 °C, the value of $\Delta S_m^\ominus$ for both the reactions

$$CaCO_3 \text{ (s)} = CaO(s) + CO_2(g)$$
$$BaCO_3 \text{ (s)} = BaO(s) + Co_2(g)$$

is 180 J K^{-1} mol^{-1}.

Calculate values of $\Delta G_m^\ominus$ for the two reactions at this temperature, using the data given in Table 8 (on p. 31). State any assumption that you make.

SAQ answers and comments

SAQ 1 NS, C_9H_{19}, $AlCl_3$, NaCl, CH and B_5H_4.
The highest common factors are four, two, two, one, six and four respectively.

SAQ 2 There are 4·20. Each formula contains twelve oxygen atoms so 1 mole $Al_2(SO_4)_3$ contains 12 moles of O. Thus 0·35 moles $Al_2(SO_4)_3$ contains $12 \times 0·35$ or 4·20 moles of O.

SAQ 3 There are $1\frac{1}{2}$. One mole of N_2H_4 contains two moles of N, $1\frac{1}{2}$ moles of N_2H_4 contain three moles of N.

SAQ 4 2·60 g chromium and 1·90 g of fluorine. One mole of CrF_2 weighs $52·0 + 2 \times 19·0$ or 90·0 g.
90·0 g CrF_2 contains 52·0 g Cr so 4·50 g CrF_2 contains
$$\frac{52·0}{90} \times 4·50 = 2·60 \text{ g Cr}.$$
Find the mass of fluorine by a similar calculation or from $4·50 - 2·60 = 1·90$ g.

SAQ 5 CH, CH_2 and CH_3 respectively.
Converting all the masses of hydrogen to be equivalent to 12 g of carbon, they become one, two and three grams respectively. Thus 12 g or one mole of carbon is combined with one, two and three moles of hydrogen atoms respectively.

SAQ 6 Under normal conditions a compound of molecular formula CH does not exist, Two well-known compounds with this empirical formula have molecular formulae C_2H_2 (acetylene) and C_6H_6 (benzene). They were described in S100, Unit 10.

SAQ 7
(a) $2H_2S + SO_2 = 3S + 2H_2O$
(b) $N_2O_4 + 2N_2H_4 = 3N_2 + 4H_2O$
(c) $4KClO_3 = 3KClO_4 + KCl$
(d) $C_5H_{12} + 8O_2 = 5CO_2 + 6H_2O$
(e) $4H_3PO_3 = 3H_3PO_4 + PH_3$
(f) $Al_4C_3 + 12H_2O = 4Al(OH)_3 + 3CH_4$

SAQ 8 0·000 3 moles.
$$0·039 3 \text{ g Xe} = \frac{0·039 3}{131} = 0·000 3 \text{ moles Xe}$$
$$0·045 6 \text{ g F}_2 = \frac{0·045 6}{38} = 0·001 2 \text{ moles F}_2$$

But in the reaction,
$$Xe + 3F_2 = XeF_6$$
1 mole Xe reacts with 3 moles of F_2 to give 1 mole of XeF_6. There is more than enough fluorine to use up all the xenon and 0·000 3 moles of XeF_6 will be formed, leaving 0·000 3 moles of unused F_2.

SAQ 9 (a) 0·288 g   (b) 0·272 g
(a) 1 mole of N_2H_4 produces 4 moles of H_2O ∴ 32 g N_2H_4 produces 72 g H_2O

$$0·128 \text{ g N}_2H_4 \text{ produces } \frac{72}{32} \times 0·128 = 0·288 \text{ g H}_2O$$

(b) 1 mole of N_2H_4 reacts with 2 moles of H_2O_2 ∴ 32 g N_2H_4 reacts with 68 g H_2O_2

$$0·128 \text{ g N}_2H_4 \text{ reacts with } \frac{68}{32} \times 0·128 = 0·272 \text{ g H}_2O_2$$

SAQ 10 The formula is PCl_5. As no chlorine was left in the filtered solution, all chlorine in the original chloride was removed as AgCl.

Molecular weight AgCl $= 107{\cdot}9 + 35{\cdot}5 = 143{\cdot}4$

$0{\cdot}286\ 8$ g AgCl contains $0{\cdot}286\ 8 \times \dfrac{35{\cdot}5}{143{\cdot}4} = 0{\cdot}071\ 0$ g chlorine.

Thus $0{\cdot}083\ 4$ g of the phosphorus chloride contains $0{\cdot}071\ 0$ g chlorine.

$(0{\cdot}083\ 4 - 0{\cdot}071\ 0)$ g or $0{\cdot}0124$ g phosphorus are combined with $0{\cdot}071\ 0$ g chlorine.

$\therefore\ 31{\cdot}0$ g phosphorus are combined with $\dfrac{0{\cdot}071\ 0}{0{\cdot}012\ 4} \times 31{\cdot}0 = 177{\cdot}5$ g chlorine.

1 mole of phosphorus is combined with 5 moles of Cl.

SAQ 11 The formula is Ag_2O.

$0{\cdot}5736$ g AgCl contains $\dfrac{107{\cdot}9}{143{\cdot}4} \times 0{\cdot}5736 = 0{\cdot}4316$ g Ag.

$0{\cdot}0320$ g oxygen are combined with $0{\cdot}4316$ g Ag.

16 g oxygen are combined with $\dfrac{0{\cdot}4316}{0{\cdot}0320} \times 16 = 215{\cdot}8$ g silver $= 2$ moles of Ag.

$\therefore$ 1 mole O is combined with 2 moles of Ag.

SAQ 12 The formula is $PtCl_4$.

$3{\cdot}73$ g chloride contains $2{\cdot}17$ g platinum.

$2{\cdot}17$ platinum are combined with $3{\cdot}73 - 2{\cdot}17 = 1{\cdot}56$ g chlorine.

$195{\cdot}1$ g platinum are combined with $\dfrac{1{\cdot}56}{2{\cdot}17} \times 195{\cdot}1 = 140{\cdot}2$ g chlorine.

1 mole of Pt is combined with $\dfrac{140{\cdot}2}{35{\cdot}45}$ moles of Cl $= 3{\cdot}96$ moles of Cl

$\therefore$ the chloride is $PtCl_4$.

SAQ 13 Np_2O_7 is a reasonable suggestion. In Figure 3, uranium falls in Group VI so neptunium would be placed in Group VII under rhenium (Re). Np_2O_7 has not yet been prepared, but recent work has yielded the anion NpO_5^{3-} in which neptunium has an oxidation number of $+7$. For *oxidation number,* see Section 6.1.6.

SAQ 14 Rhenium (Re). The highest normal oxide is M_2O_7 so the element should go in Group VII of Figure 3. The only element in this group between barium and lead is rhenium.

SAQ 15 The following diagrams are drawn without regard for shape:

SAQ 16

(a) I, zero (rule 2)
(b) U, $+3$ (rule 1)
(c) Ba, $+2$; Cl -1 (rules 5 and 7)
(d) Li, $+1$; H, -1 (rules 4 and 7)
(e) H, $+1$; S, $+4$; O, -2 (rules 3, 4 and 7)
(f) H, $+1$; Te, $+6$; O, -2 (rules 3, 4 and 7)
(g) Ga, $+3$; S, $+6$; O, -2 (rules 3, 6 and 7)
(h) H, $+1$; I, $+1$; O, -2 (rules 3, 4 and 7)
(i) I, $+7$; O, -2 (rules 3 and 7)
(j) V, $+5$; O, -2 (rules 3 and 7)

SAQ 17 In both cases, L is the typical element. In (a), the oxidation numbers of M differ by one while those of L differ by two. This suggests that M is the transition element. In (b), the oxidation numbers of L are $+1$, $+3$, $+4$ or $+5$ while those of M are $+2$, $+3$, $+4$, $+5$, $+6$ and $+7$. The ion in which L has an oxidation number of $+4$ contains two atoms which could be bonded together (see Section 6.1.8 for this qualification). If we ignore this ion, the oxidation numbers of L differ by multiples of two. If you are interested, in (a) L behaves like thallium and M like iron. In (b) L behaves like phosphorus and M like manganese.

SAQ 18

(a) $\quad Cr_2O_7^{2-} + 6Fe^{2+} + 14H^+ = 2Cr^{3+} + 6Fe^{3+} + 7H_2O$

(b) $\quad\quad\quad IO_3^- + 5I^- + 6H^+ = 3I_2 + 3H_2O$

(c) $\quad\quad 3Cu + 2NO_3^- + 8H^+ = 3Cu^{2+} + 2NO + 4H_2O$

(d) $\quad\quad 8Al + 3NO_3^- + 30H^+ = 8Al^{3+} + 3NH_4^+ + 9H_2O$

(e) $\quad 2MnO_4^- + 5H_2O_2 + 6H^+ = 2Mn^{2+} + 5O_2 + 8H_2O$

Follow the procedure in Section 6.1.7. In (a), the oxidation number of each chromium falls by *three* ($+6$ to $+3$), and the oxidation number of each iron increases by *one* ($+2$ to $+3$). Thus, three irons combine with one chromium, or 6 Fe^{2+} with one $Cr_2O_7^{2-}$. Then follow steps 5 and 6 in Section 6.1.7. Again in (c), the oxidation number of each copper increases by *two* (0 to $+2$) and the oxidation number of each nitrogen falls by *three* ($+5$ to $+2$). Thus, three coppers combine with two nitrogens, or 3Cu with $2NO_3^-$. Then follow steps 5 and 6 in Section 6.1.7. Remember that in H_2O_2 oxygen is assigned an oxidation number of -1.

SAQ 19

(i) c, (ii) b, (iii) c, (iv) d, (v) a
Disproportionation is defined in Section 6.2.8. Only the changes in the manganese oxidation states need be considered. These are:

(i) $\quad +6$ changes to $+7$ or $+4$
(ii) $\quad +6$ changes to $+4$
(iii) $\quad +3$ changes to $+2$ or $+4$
(iv) $\quad +2$ (no change)
(v) $\quad +6$ changes to $+7$.

SAQ 20 Barium hydride, BaH_2. The oxidation number of $+2$ in alkaline earth compounds means that MH_2 is the formula of the hydrides. Thus only BaH_2 and RaH_2 contain less than 2 per cent hydrogen by weight. Of these RaH_2 will be radioactive.

SAQ 21 Beryllium and magnesium. Non-existent peroxides should be unstable with respect to the normal oxide and oxygen (Section 6.2.2) and peroxide stability with respect to this reaction should increase down the alkaline earth group (Section 6.3.2). Thus the two top elements will form the least stable peroxides.

SAQ 22 $\Delta H_f^\ominus[NeCl(s)] = 1\,066$ kJ mol^{-1}. Even ignoring the relatively small $T\Delta S$ term, such a compound would be very unstable thermodynamically with respect to the reverse of the formation reaction:

$$NeCl(s) = Ne(g) + \tfrac{1}{2}Cl_2(g)$$

$\Delta H_{atm}^\ominus$ [Ne] = O because the element exists as a monatomic gas. Use the value for NaCl, -773 kJ mol^{-1}, for L_0 [NeCl]. The large positive value of $\Delta H_f^\ominus$ squares with the reluctance of neon to form compounds.

SAQ 23 The second compound in each case. See Section 6.2.3. The most acid solution will have the lowest pH.

SAQ 24 No. At 25 °C, $\Delta G_f^\ominus$ [MgH$_2$(s)] is negative so the compound is stable to the reaction

$$MgH_2(s) = Mg(s) + H_2(g)$$

No reaction between Mg and H_2 is observed because the combination of magnesium and hydrogen is very slow at room temperature, so equilibrium is not reached.

SAQ 25   The alignment of the non-spherical anions might stretch the vertical edges of the unit cell relative to the horizontal ones. Thus the unit cell might have two equal sides with the third longer. This is in fact the case.

SAQ 26   The most 'electropositive' metals are not usually obtained by reducing the oxides with carbon, although limited amounts of magnesium are made in this way (temperatures of nearly 2 000 °C are necessary). Most magnesium is made by electrolysis of the fused chloride, and one might expect the same procedure to be convenient for barium. $BaCl_2$ could be made by dissolving BaO in dilute HCl, evaporating the solution and dehydrating a hydrate of the chloride if this was obtained. In the case of barium, the hydrated chloride does not form oxygen compounds when heated.

(See also Section 5.6).

SAQ 27   The answer is caesium. The ion F^- should be smaller than the ion ClF_4^- so the stability of the compounds $MClF_4$ with respect to MF and ClF_3 at any temperature should increase down the group. Thus the preparation is most likely to succeed if francium fluoride is used. However, francium is obtainable in only trace amounts, so caesium fluoride is the practical choice.

SAQ 28

For the calcium reaction, $\Delta G_m^\ominus = -37.5$ kJ mol^{-1}.

For the barium reaction, $\Delta G_m^\ominus = +50.7$ kJ mol^{-1}.

$CaCO_3$ is therefore unstable at 1 200 K. $BaCO_3$ is stable.

Assuming that $\Delta H_m^\ominus$ does not change significantly then we can use $\Delta H_{298}^\ominus$ as $\Delta H_{1200}^\ominus$ (927 °C = 1 200 K). At this temperature, for calcium carbonate

$$\Delta G_m^\ominus = (178.5 - 1.200 \times 180.0) \text{ kJ mol}^{-1}$$

while for barium carbonate

$$\Delta G_m^\ominus = (266.7 - 1.200 \times 180.0) \text{ kJ mol}^{-1}.$$

Home Experiments for Units 6 and 7

The experiments which follow are for Units 6 and 7, although they deal with material discussed in Unit 6 only.

Experiment 1 examines some characteristics of transition metals, such as the colour and pattern of oxidation states and the formation of complex cations in water, which contrast with properties of the typical metals.

Experiment 2 deals with the chemistry of the alkaline earth metals and aluminium and Experiment 3 contains an exercise in aluminium chemistry which demonstrates Le Chatelier's Principle.

Experiment 4 involves the preparation of an aluminium compound of great practical importance.

Notes on each experiment are at the end of the Home Experiments, but do not read these until you have completed the appropriate experiment.

Use 125 × 16 mm test tubes. As before, a 'heap' is roughly that amount of material that can be balanced on the bent tip of your micro-spatula. Depths of liquid refer to 125 × 16 mm tubes.

Home experiment 1
Characteristics of transition metal compounds

The typical metals generally form few oxidation states, but those that you are familiar with, such as the alkaline earth metals and aluminium, form only one in aqueous solution. Moreover, the compounds and aqueous ions of these metals are nearly all colourless. This contrasts with the transition metals.

(a) The oxidation states of vanadium

Add two heaps of sodium metavanadate, $NaVO_3$, to a 1 cm depth of water and then add about twice the volume of dilute sulphuric acid. Stir briefly. The red crystals should nearly dissolve to give a yellow solution containing vanadium in oxidation state five, e.g. VO_3^- (aq). Now add a piece of granulated zinc and carefully heat the solution over the gas burner. Adjust the height of the test tube over the flame so that the gas evolution brings the liquid level to about three-quarters of the way up the tube. Continue heating in this way for about five minutes, occasionally removing the tube very briefly from the flame to observe the colour changes. Describe any colour changes that you see in your notebook.

(b) The oxidation states of manganese

The transition metal manganese can exist in a variety of oxidation states. By varying the conditions and the oxidizing or reducing agent, the compounds and ions listed in the Table below can be prepared.

Table Manganese compounds and ions

Oxidation state	Compound or ion	Description
+2	Mn^{2+}(aq)	colourless
+3	Mn^{3+}(aq)	claret
+4	MnO_2(s)	brown precipitate
+6	MnO_4^{2-}(aq)	green
+7	MnO_4^-(aq)	very deep purple

Perform the following experiments and in each case try to identify the product by referring to the Table:

(i) Dissolve one small crystal of potassium permanganate, $KMnO_4$, in 2 cm of water and add the solution to another test tube containing two heaps of sodium peroxide dissolved in 2 cms of cold water.

(ii) Place four pellets of sodium hydroxide, NaOH, in a test tube and *just* cover them with water. Wait until the hydroxide has dissolved, and then add *one tiny crystal* of $KMnO_4$. Observe any colour change over a period of a minute or two, giving the tube the occasional gentle shake.

(iii) Dissolve one heap of manganese sulphate, $MnSO_4$, in 2 cm of dilute sulphuric acid. When it has all dissolved add four heaps of ammonium persulphate $(NH_4)_2S_2O_8$, and one tiny crystal of silver nitrate. Stir and set aside, observing the solution at about 4-minute intervals for about 20 minutes.

(iv) Repeat (iii), but instead of a heap of manganese sulphate use one tiny crystal. Leave the solution overnight and observe its final colour.

(v) Dissolve two heaps of manganese sulphate in 1 cm of water, then add an equal volume of dilute sodium hydroxide solution. Note the colour of the precipitate at this point, and then pour off as much liquid as you can while leaving the precipitate in the test tube. Note any changes that occur during this operation, and when the tube is left to stand for a while.

(c) Complex formation by a transition metal

Dissolve four heaps of magnesium sulphate, $MgSO_4$, in 1 cm of water in one test tube and two heaps of copper sulphate, $CuSO_4$, in 1 cm of water in another. Boil the contents to dissolve the solid if necessary, but if you do this, cool under a tap afterwards.

To each test tube add an equal volume of dilute sodium hydroxide solution. Then add an equal volume of ammonia solution and stir the contents of both tubes briefly. Note the changes that occur.

Home experiment 2
The alkaline earth metals and aluminium

This section demonstrates an interesting difference between magnesium on the one hand, and sodium and aluminium on the other. It also introduces you to some important differences in the solubility of salts in the alkaline earth metal group.

(a) Reaction of magnesium and aluminium with strong alkalies

Add three or four magnesium turnings to $1\frac{1}{2}$ cm of dilute sodium hydroxide solution. Repeat, using aluminium in place of magnesium. Note what happens in both cases.

Now heat both tubes gently (do not boil), remove them from the flame and put them in your test tube rack. Observe what happens.

(b) Insoluble compounds of the alkaline earth metals

The formation of an insoluble compound, or of a particular combination of insoluble compounds, by the cation of a metal is often highly characteristic. It can therefore be used to identify the metal cation. Alternatively a solution of a cation can be used to identify an unknown anion.

The alkaline earth metal cations form compounds whose solubilities usually either increase or decrease steadily down the group from magnesium to radium. Given this piece of information, try to use the experiments which follow to decide whether the radium salt of the four anions used below is likely to be the most soluble or the least soluble of the series of alkaline earth metal compounds formed with that anion. In at least one case the experiment cannot provide an answer.

Dissolve six heaps of strontium dichloride, $SrCl_2$, in 4 cm of water in one test tube and six heaps of magnesium sulphate, $MgSO_4$, in 4 cms of water in another. Divide the strontium solution equally among four test tubes and the magnesium solution equally among another four. You now have four pairs, each pair consisting of one magnesium and one strontium solution.

(i)   To one pair, add equal volumes of sodium carbonate solution made up by dissolving six heaps of the carbonate in 4 cm of water.

(ii)   To a second pair, add equal volumes of dilute sulphuric acid, H_2SO_4.

(iii)   To a third pair, add equal volumes of dilute sodium hydroxide, NaOH.

(iv)   To the fourth pair, add equal volumes of dilute hydrochloric acid, HCl.

Home experiment 3
Another example of Le Chatelier's Principle

Dissolve four heaps of aluminium sulphate in 4 cm of water, and divide the solution equally between three test tubes. Now dissolve four heaps of disodium hydrogen phosphate, Na_2HPO_4, in 4 cm of water. This solution acts as a source of phosphate ions, PO_4^{3-}, and precipitates cations which form insoluble phosphates. Add one-third of the solution to the first test tube of $Al_2(SO_4)_3$ solution, one-third to the second, and one-third to the third.

> Is aluminium phosphate soluble in water? Write down its formula.

Now add an equal volume of dilute sulphuric acid to the first test tube, an equal volume of dilute acetic acid to the second and an equal volume of dilute sodium hydroxide to the third.

Note down your observations. Can you account for what you see?

Home experiment 4
The preparation of a compound; potash alum

Place just over half your sample of potassium hydroxide, KOH, in a large boiling tube, and add 10 cm^3 of water. Allow the solid to dissolve.

Now take about one-third of the initial quantity of aluminium turnings supplied with the kit, and add it little by little to the solution. This regulated addition is recommended because the reaction can become quite violent, and you do not want the corrosive solution to bubble out of the tube. If this looks like happening, cooling under a tap will help. At first you will have to keep a watch on the reaction, but eventually it will moderate so much that you will be able to add the remaining aluminium and leave it until reaction has ceased. This should take about 45 minutes. Do not worry about any solid residue in the tube.

Now gradually add about 20 cm³ of dilute sulphuric acid, stirring the solution. A white precipitate should appear. Transfer as much of the contents of the test tube as you can to a 100 cm³ beaker, then add a further 15 cm³ of sulphuric acid to the test tube, stir and quickly add it to the beaker as well. Stir again. Test the solution with a piece of blue litmus paper to see if it is acid, and if not, add H_2SO_4, little by little, until it is.

Now heat the beaker on a tripod and gauze, and allow the solution to boil gently for a minute or two during which most of the solid should dissolve.

While this is happening, fold a filter paper in half to form a semi-circle, then fold the semi-circle in half to form a quarter-circle. Now open out the rim of the quarter-circle to form a cone so that there are three thicknesses of paper on one side and one on the other. Insert this cone in the filter funnel and dampen it with a little water to stick it in place. Support the funnel on a retort stand and place a clean dry 250 cm³ beaker below to collect the filtered liquid (the filtrate).

Pick up the hot beaker in a thick cloth or wad of tissue paper in one hand, supporting the underside of the beaker with another part of the cloth held in the other hand. Carefully pour the hot liquid into the funnel until it rises to about ½ cm below the top of the filter paper. The liquid should filter quite quickly and you should top up the level of liquid frequently, reheating the unadded liquid in the meanwhile if you are after a better yield.

When all the liquid has filtered, cover the beaker of filtrate with a watch glass or a piece of paper, and leave it overnight in a cold place (preferably in a refrigerator).

Potash alum crystallizes out as the solution cools. Next day, pour off as much liquid from the crystals as you can into a boiling tube and transfer the mass of wet crystals that remain to a filter paper pad made up of four or five papers. Press the crystals with another pad of four or five papers to remove as much moisture as possible and then leave them in a warm place to dry out. Inspect them. You should find that square or triangular faces are common.

Crystals can be made to grow in a saturated solution if left in a cold place for a long time. This can be done by picking out a good crystal, looping it in cotton and hanging it from a matchstick in the boiling tube of liquid that you poured off, to which excess solid has been added.

Potash alum has a rather complicated formula, sometimes written $K_2SO_4 \cdot Al_2(SO_4) \cdot 24H_2O$ and sometimes $KAl(SO_4)_2 \cdot 12H_2O$. In water, it gives K^+, Al^{3+} and SO_4^{2-} ions. It is the most important member of a family of compounds called *alums* in which K^+ can be replaced by Na^+, Rb^+ or Cs^+ and Al^{3+} by certain other tripositive ions. You can see that potash alum contains many molecules of water of crystallization and, as these are driven off by heat, the compound is used for fire-proofing fabrics. The most important use however is a very old one – in dyeing. Some points of interest are described in the notes on this experiment.

Notes on the Experiments

Experiment 1(a)

Usually the sequence of colours observed is yellow, green, blue, green and, finally, a steel blue or lavender colour. The first green colour observed is a mixture of the starting material, VO_3^-(aq), with the blue colour of the ion VO^{2+} (oxidation state $+4$). This is then reduced to the green V^{3+}(aq) which, with some difficulty, can usually be reduced further to the lavender V^{2+}(aq). Note that the oxidation states differ by one (see Section 6.1.8).

Experiment 1(b)

(i) Mn^{2+} is formed. The reducing agent is hydrogen peroxide formed by the sodium peroxide (see Section 6.1.2.).

(ii)   In concentrated alkali, like the solution used here, MnO_4^- is reduced by water to give the green MnO_4^{2-}. This decomposes unless the solution is very alkaline.

(iii)   Experiments at the Open University gave the rose colour of a very dilute solution of Mn^{3+} after 5 minutes. It so happens that this is unstable in water except at very low concentrations, and as its concentration builds up, it disproportionates giving MnO_2 and Mn^{2+}.

$$2Mn^{3+}(aq) + 2H_2O = MnO_2 + Mn^{2+} + 4H^+$$

The brown colour of MnO_2 was visible after 15 minutes.

(iv)   You should obtain purple MnO_4^-. Mn^{3+} can be oxidized to MnO_4^-, but MnO_2 is more difficult, so if a high concentration of Mn^{2+} is used to begin with, disproportionation of Mn^{3+} tends to stop the reaction at MnO_2 as in (iii). With a low concentration of Mn^{2+}, disproportionation of Mn^{3+} does not occur and MnO_4^- is formed. In both (iii) and (iv) the oxidizing agent is the persulphate anion, $S_2O_8^{2-}$; the Ag^+ ion acts as a *catalyst*: $5S_2O_8^{2-} + 2Mn^{2+} + 8H_2O = 2MnO_4^- + 10SO_4^{2-} + 16H^+$.

(v)   In alkaline solution, a flesh coloured precipitate of $Mn(OH)_2$ is formed at first. Atmospheric oxygen gradually oxidizes this to MnO_2.

Experiment 1(c)

Both hydroxides are insoluble, but that of copper dissolves in ammonia to form the deep blue complex ion $Cu(NH_3)_4^{2+}$. This illustrates the property of transition metal ions discussed in Section 6.2.6.

Notice the action of Le Chatelier's principle. For both hydroxides the equilibrium position in

$$M(OH)_2(s) = M^{2+}(aq) + 2OH^-(aq)$$

lies to the left in water. But when ammonia is added, $Cu^{2+}(aq)$ (but not $Mg^{2+}(aq)$) forms a complex cation, the equilibrium above is displaced to the right, and $Cu(OH)_2$ dissolves to supply more $Cu^{2+}(aq)$.

Experiment 2(a)

The reaction of aluminium becomes quite violent while that of magnesium is very slow.

Experiment 2(b)

(i)   Both carbonates are insoluble, so no conclusion is possible.

(ii) $SrSO_4$ is insoluble, $MgSO_4$ is not. Thus $RaSO_4$ should be insoluble, and at the same time, the most insoluble sulphate. $BaSO_4$ is also insoluble, and barium chloride solution is used to test solutions for sulphate.

(iii) $Mg(OH)_2$ is insoluble and $Sr(OH)_2$ is not, so $Ra(OH)_2$ should be soluble.

(iv)   Both chlorides are soluble so no conclusion is possible. In fact, the solubility of the chlorides decreases down the group, and radium chloride is the least soluble. By using the lower solubility of radium chloride, Marie Curie was able to separate radium from barium after she had first used the insolubility of their sulphates ((ii) above) to separate them together from the bulk of the material. From two truckloads of Pitchblende residues supplied from an Austrian mine, and containing radium in minute concentrations she eventually obtained 0.12 g of pure radium dichloride, $RaCl_2$, a most remarkable experimental achievement.

Experiment 3

Aluminium phosphate, which is insoluble in water, has the formula $AlPO_4$.

As stated in the text, the aqueous ion in a solution of disodium hydrogen phosphate, $HPO_4^{2-}(aq)$ can act as a source of phosphate ions although the equilibrium lies well to the left. ($HPO_4^{2-}(aq)$ is a very weak acid.)

$$HPO_4^{2-}(aq) \rightleftharpoons H^+(aq) + PO_4^{3-}(aq) \tag{1}$$

In the presence of $Al^{3+}(aq)$ the insoluble phosphate is formed and the equilibrium is displaced to the right as the phosphate ions are removed from solution.

A second equilibrium

$$AlPO_4(s) \rightleftharpoons Al^{3+}(aq) + PO_4^{3-}(aq) \tag{2}$$

is then attained.

A strong acid displaces equilibrium 1 extensively to the left, and, consequently, 2 to the right. Acetic acid, however, does not cause a sufficient displacement for the phosphate to dissolve.

Sodium hydroxide dissolves $AlPO_4$ because the Al^{3+} ion is removed by the reaction

$$Al^{3+}(aq) + 4OH^-(aq) = Al(OH)_4^-(aq)$$

and equilibrium 2 is displaced to the right.

Experiment 4

Alums containing the Al^{3+} ion are formed not only by K^+, but also by the cations Na^+, Rb^+, Cs^+ and NH_4^+. The ammonium and potassium alums in particular have a long history of domestic and artistic uses. Both solutions of simple aluminium salts and those of alums contain aluminium ions, and natural fibres like wool, which were soaked in these solutions, were found to absorb dyes very readily. Substances producing this effect are called mordants. The prime importance of alums stemmed from their natural occurrence. Ordinary clays contain, among other things, potassium ions, and aluminium and oxygen atoms bonded together. In the vicinity of volcanoes, sulphuric acid vapour acts upon the clay to produce either potash alum, $KAl(SO_4)_2 \cdot 12H_2O$, or more complex basic sulphates such as 'alunite'. $K_2SO_4 \cdot Al_2(SO_4)_3 \cdot 4Al(OH)_3$ from which an alum solution is easily obtained by adding water.

Until the fifteenth century, the European dyeing industry obtained much of its alum from acid volcanic regions in North Africa and what is now Central Turkey. The bulk of the trade fell into the hands of Italian City States, such as Venice and Genoa, which lay close to the Mediterranean Sea Routes in a position well suited to the distribution of goods into Central Europe. The trade was so profitable that when new sources of alum were discovered at Tolfa near Rome in the Papal States in the 15th century, Pope Paul II tried to set up a monopoly by threatening those who purchased alum from the Turks with excommunication. After the consequent price rise, the usual problems of enforcement were experienced, especially with the British, of whom one of the Papal emissaries remarked: 'In the morning they are as devout as angels, but after dinner they are like devils, seeking to throw the Pope's messenger into the sea.'

Unit 7 Classical Bonding Theories

Contents

Objectives

1   Define in your own words, recognize valid definitions of, and use in a correct context the terms or expressions in Table A.

2   Draw diagrams or coordinate plans of the unit cells of the caesium chloride, sodium chloride, zinc blende and fluorite structures.

3   Given the unit cell of a compound, calculate the number of molecules in it.

4   Rationalize the estimation of values for the radii of the iodide and other ions of the alkali metal halides using the internuclear distances and a hard sphere model.

5   Account for differences in the coordination number of cations in alkali metal halides or in alkaline earth metal halides in terms of the relative sizes of ions.

6   Account for the differences in the stabilities of certain alkali metal and alkaline earth metal compounds by using thermodynamic cycles and simple ideas about the variation of the lattice energies of ionic compounds with ion size.

7   Describe the coordination of the atoms in the rutile, cadmium chloride and cadmium iodide structures.

8   Use valence shell repulsion theory to predict shapes for simple molecules of the typical elements.

9   Cite structural features to support the belief that the chlorides become more covalent across the third period of typical elements.

10   Use the concept of electronegativity or ion polarization to rationalize different degrees of ionic character in the halides of different elements.

11   Draw resonance structures, given the molecular geometry of highly symmetrical molecules and other data on bond lengths.

12   Describe the reasoning which led to the preparation of the first noble gas compound (see the TV programme for this Unit).

Table A

List of scientific terms, concepts and principles used in Unit 7

Introduced in a previous Unit	Unit Section No.	Developed in this Unit	Page No.
	S100*	caesium chloride structure	6
potential energy	4.4.3	dative bond	17
sodium chloride structure	8.4.6	fluorite structure	14
valence shell repulsion		ionic radius	8
theory	10.4.1	lattice energy	10
		layer structures	14
	S25-	polarizability	23
Born-Haber cycle	5.8	polarization of ions	23
coordination number	3.4	polarizing power	23
electronegativity	5, App. 1	resonance structures	24
enthalpy change	3, 4	rutile structure	14
lattice energy	5, 6	valence shell repulsion theory	18
Lewis theory	3.3	zinc blende structure	6.
microwave spectroscopy	1.2		
unit cell	2.4		
X-ray diffraction	2		

* The Open University (1971) S100 *Science: A Foundation Course*, The Open University Press.

Study guide

As the title of our Course suggests, the subjects of structure and bonding are closely related, and in this Unit we examine the connection more closely by using the structures of some compounds of the typical elements to evaluate the elementary bonding theories inherited from the Foundation Course in Science (S100).

The Unit can be conveniently divided into parts. In Sections 7.1–7.1.6, we develop and test a model in which compounds normally classified as 'ionic' are regarded as composed of charged spheres whose mode of packing affects structure. Then, in Sections 7.2 and 7.2.1, we examine some thermodynamic implications of such a model.

In Sections 7.3–7.4.3 we look at a selection of compounds, some of whose structures show marked deviations from the characteristics expected of ionic compounds. This selection includes covalent molecules whose shapes can be treated by valence shell repulsion theory.

The structural details disclose certain weaknesses in our elementary theories; some of these weaknesses are dealt with specifically in Sections 7.5 and 7.6, and a more general discussion is given in the Summary in Section 7.7.

7.1 Introduction

In Unit 6, we studied the chemistry of the alkali metals, the alkaline earth metals and aluminium. To judge by their physical and chemical properties, most of the solid compounds that were described could be classified as 'ionic', but in this Unit we shall see that this impression is only partly reinforced by an examination of their structures. This difficulty, among others, leads us to an appreciation of the need for the introduction of new bonding theories.

The ideas discussed in this Unit were developed after the publication of G. N. Lewis' famous paper of 1916 in which he introduced the idea of the two-electron bond. They represent attempts to develop Lewis' models of ionic and covalent bonding so that these models are more consistent with the observed structural and physical properties of ionic and covalent compounds. Developments of the ionic model are dealt with in Sections 7.1.5, 7.1.6 and 7.2, and developments in the covalent bonding model in Sections 7.4, 7.4.1 and 7.4.2. A second problem that we shall be concerned with was one of Lewis' main preoccupations: how can the separate concepts of ionic and covalent bonding be welded together into a coherent and satisfying whole? Partly successful attempts to do this are discussed in Sections 7.3.4 and 7.4.3.

Important background to this Unit is provided by the third radio programme of this Course, which is designated S25-02 and called *150 Years of Bonding Theory*.

In this Unit we begin by discussing some common crystal structures, and then consider the ways in which real compounds are distributed among these possible structures.

The set book is R. C. Evans (1966), *An Introduction to Crystal Chemistry*, 2nd ed., Cambridge University Press, hereafter referred to as Evans.

7.1.1 The sodium chloride structure

Read Evans, Section 3.04, ignoring the footnote, and then try to answer the following questions:

> The coordination number of sodium is the same as the coordination number of chlorine. Why is this?

If the structure is of the type MX, a crystal must contain equal numbers of sodiums and chlorines. Thus if a sodium is surrounded by six chlorines, a chlorine must be surrounded by six sodiums.

> Suppose that the sodiums are represented by the solid black circles in Evans, Figure 3.01. How many sodiums does the unit cell contain?

The answer is four. There are twelve sodiums at the mid-points of the edges of the cube. You can probably see that these edges cut one-quarter of each sodium sphere into the unit cell. Alternatively, when unit cells are stacked together, each edge, and therefore each sodium at its mid-point, is shared between four unit cells. Thus 12 sodiums on the edges make up $12 \times \frac{1}{4} = 3$ whole sodiums. In addition, there is one complete sodium at the centre of the unit cell and this makes four in all.

> If the chlorines are represented by the open circles, how many chlorines does the unit cell contain?

Again the answer is four. There are eight chlorines at the eight corners of the cube, and six at the centres of the six faces. Each corner is shared between eight unit cells, and cuts one-eighth of a chlorine into each. Each face is shared between two unit cells, and cuts one-half of the chlorine at the centre of each face into each unit cell. Thus the total number of chlorines in each cell is $8 \times \frac{1}{8} + 6 \times \frac{1}{2} = 4$. Notice that the numbers of the chlorines and sodiums in each cell are equal. This must be so because the formula of the compound, NaCl, has the same property. As the text states, we cannot pick out discrete molecules of sodium chloride. However, each unit cell contains four sodiums and four chlorines, and it is customary to express this by saying that it contains four molecules of sodium chloride.

7.1.2 The caesium chloride and zinc blende structures

Read Evans, Sections 3.05 and 3.06.

Representing structures in three dimensions is not too easy, and it is a good idea to master the trick of drawing two-dimensional plans of them. In Evans, Figure 3.03, x, y and z axes are marked for the caesium chloride unit cell. Looking down the z axis, we can represent the unit cell by a plan which is shown in Figure 1. We shall call such a diagram *a coordinate plan*.

Positions of atoms or ions in crystal structures are decribed by x, y and z coordinates and these coordinates are expressed as fractions of the total distance that the x, y and z axes take to traverse the unit cell. In diagrams like that in Figure 1, the x and y coordinates are not recorded because they can simply be represented by their position in the plane of the paper. Moreover, the z coordinate is recorded against an atom or ion only if it is fractional. This is because an atom or ion with $z = 0$ must, by the repetitive qualities of the unit cell, occur again at $z = 1$, so the open circles at the corner of the square in Figure 1 represent atoms or ions of that type at $z = 0$ and $z = 1$.

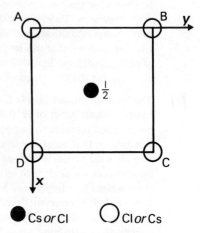

Cs *or* Cl ⬤   **Cl *or* Cs** ◯

Figure 1 Plan of the CsCl structure.

> **SAQ 1 (Objectives 2 and 3)**   How many molecules of caesium chloride does the unit cell contain?

Now write down the coordinates ($x\ y\ z$) of the nine particles in Evans, Figure 3.03, with respect to the origin of the axes.

> **SAQ 2 (Objectives 2 and 3)**   Draw a coordinate plan of the zinc blende structure shown in Evans, Figure 3.04.

> **SAQ 3 (Objectives 2 and 3)**   How many molecules are there in the zinc blende unit cell?

The eight open circles are at (000), (100), (010), (001), (110), (101), (011), (111). The single filled circle is at $(\frac{1}{2}\frac{1}{2}\frac{1}{2})$.

7.1.3 Internuclear distances in ionic compounds

As you know from Unit 2, the shortest distance between different atoms in a solid, such as sodium chloride or caesium chloride, can be obtained by X-ray crystallography. From a list of internuclear distances for the alkali metal halides, an important observation can be made.

Now read Evans, Section 3.07.

In this Course, which uses SI Units, we give internuclear distances in picometres (pm) (1 pm $= 10^{-12}$ metres) whereas in Evans these distances are given in angstrom units (Å) (1 Å $= 10^{-10}$ metres). Remember, therefore, that to convert figures given in Evans to figures used in this Unit, you must multiply by $10^{-10}/10^{-12}$, that is, by one hundred.

To summarize what you have just read, we shall select the internuclear distances for the halides of sodium and potassium from Evans, Table 3.01. The eight halides all have the sodium chloride structure at room temperature and pressure.

Table 1 Internuclear distances in the halides of sodium and potassium (picometres)

	F	Cl	Br	I
Na	231	281	298	323
K	266	314	329	353

By comparing the upper of the two numbers in any column with the lower, you can find out what happens to the internuclear distance when sodium in a sodium halide is replaced by potassium.

What does happen?

In all four instances, the internuclear distance increases. What is more, all the increases are nearly the same; they lie in the range 33 ± 3 pm.

An increase by the same amount is just the result that we would expect if the ions in the crystals were hard spheres of a fixed size in contact with one another, the radius of K^+ being about 33 pm greater than that of Na^+. Thus, the variation in the internuclear distance is nearly satisfied by a model which takes ions to be hard spheres. In fact, you can see that within a row or column in Evans, Table 3.01, values of Δ are nearly constant. This shows that the conclusions that we have made from Table 1 apply to the more extensive data of Evans, Table 3.01.

If we want to regard the compounds in Evans, Table 3.01 as composed of ions, then we can build the property that we have just described into our ionic model by assigning a radius to each of the five alkali metal cations, and a radius to each of the four halide anions so that when one of the five is added to one of the four, something very close to the internuclear distance in the appropriate alkali metal halide is obtained. This means that the 20 distances in Evans, Table 3.01, can be obtained from only 9 radii, although the variation of Δ along the columns or ions shows that the internuclear distance obtained in this way will usually not be *exactly* equal to its experimental value.

To obtain our nine radii we have to do one vital thing: assign one radius to a particular ion, say sodium. By subtracting the radius of this sodium ion from the internuclear distances in the sodium halides, we can then obtain radii for F^-, Cl^-, Br^- and I^-, and these four anion radii can be used to derive radii for Li^+, K^+, Rb^+ and Cs^+.

A method of assigning the first key radius is described in Section 7.1.5, but before we consider it, we must look more closely at some internuclear distances.

7.1.4 Preliminary investigation of ion size

Let us first try to see how the internuclear distances in ionic crystals vary with the charges that we assign to the ions of which they are composed. The ions Na^+, Mg^{2+}, O^{2-} and F^- all have the same number of electrons. Now in both NaF and MgF_2, the metal ions are surrounded by six fluorines at the corners of an octahedron (the detailed structure of MgF_2 is described later in Section 7.3.2). The internuclear distances are given in Table 2.

Table 2   Internuclear distances in NaF, MgF₂ and MgO (picometres)

NaF	231	MgF_2	201
MgF_2	201	MgO	210

The shortening as we move from NaF to MgF_2 suggests that, although the cations have the same number of electrons, the radius of Mg^{2+} is about 30 pm less than that of Na^+. However, when we consider anions and compare MgF_2 with MgO, the distances suggest that the doubly charged O^{2-} ion is slightly larger than F^-.

Many more examples similar to those given in Table 2 could be cited in support of the idea that the radii of cations decrease as the positive charge increases, while the radii of anions, if anything, increase slightly as the negative charge increases.

Can you relate this to the electronic structure of atoms and ions?

One could assume that the size of ions is dependent on the spatial distribution of the electrons around the nucleus. As cations increase their charge by loss of electrons, and this eases the repulsive forces between the electrons that are left, these electrons collapse into a smaller space around the nucleus. Conversely, acquisition of electrons by a negative ion increases the repulsive forces which are eased by expansion of the electron distribution.

Both this explanation, and a straightforward extension of the empirical observations to include atoms, lead to the idea that cations are considerably smaller than their parent atoms while anions are larger. Thus if one assumes the existence of ions in compounds, one is led to the conclusion that, in general, cations tend to be considerably smaller than anions.

7.1.5 Ionic radii

As you learnt in Unit 2, the first structure-determination by X-ray diffraction was carried out by W. H. Bragg and his son, the late Sir Lawrence Bragg, when they solved the structure of sodium chloride in 1913. The study of other structures soon disclosed the additive quality of internuclear distance (described in Section 7.1.3) and, when G. N. Lewis' theories gave added weight to the idea that compounds such as the alkali metal halides were ionic, it was natural to try to assign radii to individual ions. As we pointed out in Section 7.1.3, this involves a key initial assignment of a radius to one particular ion. The method described in this Section is based upon one of the earliest ways of doing this; it was first used by a German, A. Landé, in 1920.

To follow the method through, we must return to those alkali metal halides which have the sodium chloride structure. The alkali metal cations are surrounded octahedrally by six halide anions as in Evans, Figure 3.02. The horizontal plane through the centre of the octahedron includes the central cation and four of the six surrounding anions. Now, using our hard-sphere model, we expect the anions to pack tightly against the cations as in Figure 2.

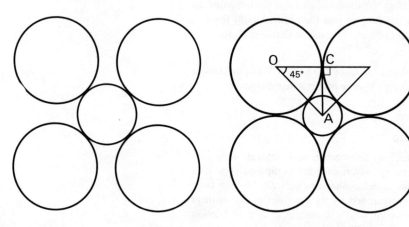

Figure 2 Anions packed around a cation on a horizontal plane.

Figure 3 Anion–anion contact.

However, if the size of the cation sphere is steadily decreased, there will come a time when the anions will touch one another as in Figure 3.

If we could detect this situation in some way, then we could calculate the anion radius in that structure.

How do you think this could be done?

The anion radius in Figure 3 is OC and $\widehat{COA} = 45°$. Thus

$$OC = AO \cos 45°$$

$$= AO/\sqrt{2}$$

where AO is the internuclear distance that we know.

Unfortunately it has not proved possible to detect experimentally a situation corresponding to anion–anion contact with complete confidence, but it is clearly most likely to occur when cations are very small, and anions are very large.

Now, the increase in internuclear distance along the rows of Evans, Table 3.01, suggests that cation size increases as we move from Li^+ to Cs^+, while the decrease down the columns suggests that anion size increases from F^- to I^-. Thus in the context of our hard-sphere model, anion–anion contact is most likely in LiI.

SAQ 4 (Objective 4) Estimate a value for the radius of the iodide ion using the data in Evans, Table 3.01.

Your answer to SAQ 4 should have been 2.12 Å or 212 pm. Now we can use this radius to get values for the radii of all the other ions involved in the alkali metal halides. However, we have to face two problems.

8

1 The slight variations in the values of Δ in Evans, Table 3.01, as we move along a row or down a column means that the internuclear distances are not precisely additive. This means that we get different values for a particular ion if we use different halides.

SAQ 5 (Objective 4)   Calculate a radius for F^- from the data in Evans, Table 3.01, for NaI and NaF. Repeat the calculation using RbI and RbF.

From SAQ 5 you should have got two different values for the fluoride ion radius, so our first problem is to juggle statistically with the internuclear distances and radii so that we get a set of radii which give the best fit.

2 Our calculations place a heavy reliance on the radius of I^- deduced from LiI. If we consider other compounds where we think anion–anion contact might occur, we might obtain single radii which generate a somewhat different set of ionic radii from those obtained using an iodide radius of 212 pm. Again, this is a problem which can be overcome by a statistical analysis of a sufficiently wide range of internuclear distances.

The radii in Table 3 have been obtained by the kind of calculation described in this Section and, at the same time, statistical steps have been taken to counter problems 1 and 2. Notice that the final value for the iodide radius is slightly different from 212 pm.

Table 3   Ionic radii (picometres)

Li^+	68	Mg^{2+}	68	Al^{3+}	.50	H^-	148
Na^+	100	Ca^{2+}	99	Fe^{2+}	75	F^-	133
K^+	133	Sr^{2+}	116	Fe^{3+}	61	Cl^-	182
Rb^+	147	Ba^{2+}	134	Sc^{3+}	69	Br^-	198
Cs^+	168	Ra^{2+}	143	Y^{3+}	86	I^-	220
		Zn^{2+}	70	La^{3+}	102	O^{2-}	142
		Cd^{2+}	97			S^{2-}	184
		Hg^{2+}	103				
		Pb^{2+}	121				

As we stated at the beginning of this Section, the basic assumption on which this method of deriving ionic radii is based (anion–anion contact in selected compounds) was first used by Landé in 1920. Since that time, other ways of deriving ionic radii have been devised, but we shall not consider them here. We merely note that they give similar results to those in Table 3, and that of all the methods, the one used here is most closely related to the behaviour expected of hard spheres.

The most important trends to note from Table 3 are the increases in the radii of similarly charged ions as one moves down a group of the Periodic Table. Compare, for example, the values for the alkali metal cations and for the alkaline earth cations. Again, the radii of the halide anions show the same trend.

Note too the radii of the ions Na^+, Mg^{2+} and Al^{3+}. The parent atoms are adjacent in the Periodic Table, and the three ions have the same number of electrons. You can see from Table 3 that the ionic radii decrease as we move from sodium to aluminium. In this case, the number of electrons remains constant, but the atomic number or nuclear charge increases. Thus, as we move from sodium to aluminium, the electrons are pulled in towards the nucleus and the radii decrease. This tendency for cations to become smaller as their charge is increased was rationalized also for the case of cations of the same element, in Section 7.1.4.

We shall now consider two applications of ionic radii. In Section 7.1.6, we relate the relative stabilities of possible crystal structures to the conditions that affect the close-packing of differently sized spheres. Then, in Section 7.2, we examine the relation between ionic radii and the bonding energy of an assembly of ions.

7.1.6 Crystal structure and ion size

In an ionic compound, we argue that the forces between the positive and negative particles hold the structure together. At first sight therefore, we might expect the structure which puts any ion into contact with the largest number of oppositely charged ions to be the one of lowest energy. If our reasoning is correct, then this structure should be the most common.

Of the three MX structures that we have considered in Sections 7.1.1 and 7.1.2, we might expect the CsCl structure to be the most favoured one because each ion packs the largest number of oppositely charged ions around it. However, all the alkali metal halides possess a sodium chloride lattice except for CsCl, CsBr and CsI which have the CsCl structure. This shows that the number of cation–anion contacts is not the only factor which affects the stabilities of different structures.

We can use our hard-sphere ionic model to account for this. Suppose we have an ionic compound in which each ion is surrounded by a large number of oppositely charged ions. It is convenient to use the sodium chloride structure as an example. As we pointed out in Section 7.1.5, the horizontal plane of the octahedron of anions around each cation includes the central cation and four of the six surrounding anions as in Figure 2. If the size of the cation is steadily decreased, there will come a time when the anions will touch one another as in Figure 3. Once this situation is reached, then according to our model, any further decrease in cation size must break the contact between the cation and its surrounding anions. This means that a further decrease in cation size will no longer decrease the distance between the centres of oppositely charged ions or, therefore, increase the attraction between the ions in the lattice. A structure with a lower coordination number, such as the zinc blende structure, must therefore be adopted if full cation–anion contact is to be maintained.

To summarize this Section, let us reverse the progression in coordination number. If we consider an ionic structure in which cations are surrounded by a relatively low number of anions, say four, then if the cation radius, r_+ is increased, there will come a time when the ratio r_+/r_- is large enough for *six* anions to be packed around the cation without anion–anion contact occurring. The structure of higher coordination number will then be the more stable. Further increases in cation size will bring the ratio r_+/r_- to a value where *eight* anions can be packed around the cation without any anion–anion contact and, if our arguments are applicable, a new structure will be observed. One important specific conclusion here is that as the ionic radii of the alkali metal and the alkaline earth metal cations increase down the group, the coordination numbers in the compounds formed with a particular anion should also increase down the group. Thus if we consider the alkali metal chlorides, the lithium, sodium, potassium and rubidium compounds have the six-coordinate NaCl structure while the caesium compound has the eight-coordinate CsCl structure. As pointed out in Evans, Section 3.05, this is also true of the alkali metal bromides and iodides. We shall meet further examples among compounds of the alkaline earth metals in Section 7.3 of this Unit.

Our arguments in this Section have been purely qualitative, but it may have occurred to you that, since we have actual values for ionic radii, we could use a quantitative approach. This is indeed possible and, if you are interested, we show you how this could be done in Appendix 1 (Black). This shows that at a quantitative level the attempt to use the hard-sphere ionic model to account for changes in crystal structures is only partly successful.

7.2 The lattice energy

You were introduced to the lattice energy of a solid metal halide when the Born–Haber cycle was discussed in Unit 5, Section 5.8. Moreover, in Unit 6, Sections 6.2.7 and 6.2.8, we used the idea that, if it was justifiable to regard compounds as composed of ions, then two compounds of the same formula type containing ions of similar sizes were likely to have similar lattice energies. This enabled us to estimate the standard enthalpies of formation of some unknown solids such as $NaCl_2$, MgCl and AlCl and to show that the estimates were consistent with the non-existence of such compounds.

As we have now obtained a set of ionic radii which we hope provides something approaching a quantitative measure of ion size, our next task will be to examine possible relationships between lattice energies and ionic radii.

The lattice energy of a compound L_0, is the internal energy change at 0 K, ΔU_0, when defined gaseous ions are brought together from an infinite distance apart to form the solid compound. Let us begin by considering just two oppositely charged ions.

Suppose two oppositely charged spheres are brought together from an infinite distance apart. Then according to S100, Unit 4, Section 4.4.3, their potential energy is lowered by the amount

$$\frac{z_+ z_- e^2}{4\pi\epsilon_0 r}$$

where z_+ and z_- are the numbers of positive and the numbers of negative charges carried by the spheres, and r is the distance between the centres of the spheres.

Now suppose we have two doubly charged ions, M^{2+} and X^{2-} in contact as in Figure 4, and that these have been brought together from an infinite distance apart.

As the spheres are in contact at the close of the operation

$$r = r\,(M^{2+}) + r\,(X^{2-})$$

If we keep the charges fixed and just vary the ionic radii, the potential energy change when the spheres are brought together is given by ΔE where

$$\Delta E = -\frac{C'}{r\,(M^{2+}) + r\,(X^{2-})} \tag{1}$$

and C' is a positive constant equal to

$$\frac{4e^2}{4\pi\epsilon_0} \text{ or } \frac{e^2}{\pi\epsilon_0}$$

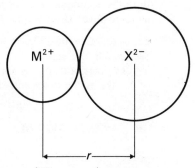

Figure 4 Cation and anion in contact.

Now suppose the ions in the solid oxide of an alkaline earth metal are brought together from an infinite distance from one another. The process is

$$M^{2+}(g) + O^{2-}(g) = MO(s)$$

and the change in potential energy is, in fact, very nearly equal to the lattice energy L_0.

We cannot expect L_0 to be equal to ΔE in equation 1 because ΔE refers to only two ions and, as the oxides of alkaline earth metals all have the sodium chloride structure, in the solid oxide each ion is surrounded by *six* of opposite charge. However there is obviously a relation between L_0 and ΔE because whether two oppositely charged ions are brought together, or whether a mole of oppositely charged ions is brought together to form a mole of solid oxide, the decrease in energy is larger the smaller the sum of the radii of the two kinds of ion. In fact, it turns out that for a series of compounds which have the same formula type and which contain ions of the same charge (e.g. the alkaline earth metal oxides, etc.) it is quite a good approximation to put

$$L_0 = -\frac{C}{r_+ + r_-} \tag{2}$$

where r_+ is the radius of the cation, r_- is the radius of the anion, and C is a positive constant which is different from C' in equation 1. We shall find that this equation is very useful in its general form. As we have said, C is nearly constant for compounds which have the same formula type, and which contain ions of the same charge. In Section 6.2.7 and 6.2.8 we assumed, for example, that the lattice energies of $NaCl_2$ and $MgCl_2$ were identical because the sizes of the cations were likely to be similar. You can now see that such assumptions are a consequence of equation 2.

SAQ 6 (Objective 6) Examine the groups of compounds (i)–(iv) below. In which groups is the value of C in equation 2 the same for all compounds in the group?

(i) NaCl, KCl, MgO, SrO

(ii) MgF_2, $CaCl_2$, CaF_2, $SrCl_2$, $BaBr_2$

(iii) NaF, NaCl, NaBr, NaI

(iv) MgF_2, NaF, KF, CaF_2

7.2.1 Lattice energies and thermal stability

To demonstrate the value of equation 2, we shall re-examine the thermal stabilities of the alkaline earth carbonates. In Unit 6, Section 6.3.1, we established a connection between the decomposition temperatures, and the values of $\Delta H_m^\ominus$ at 298.15 K for the decomposition reaction

$$MCO_3(s) = MO(s) + CO_2(g) \qquad (3)$$

Both the decomposition temperatures and the value of $\Delta H_m^\ominus$ increased (became more positive) as we moved down Group II from beryllium to barium.

Let us examine this problem by using the method that was introduced in Unit 5, Section 5.8. We can carry out reaction 3 in convenient steps of our own choosing and create a thermodynamic cycle. Consider the step by step decomposition of MCO_3 by the lower anti-clockwise route.

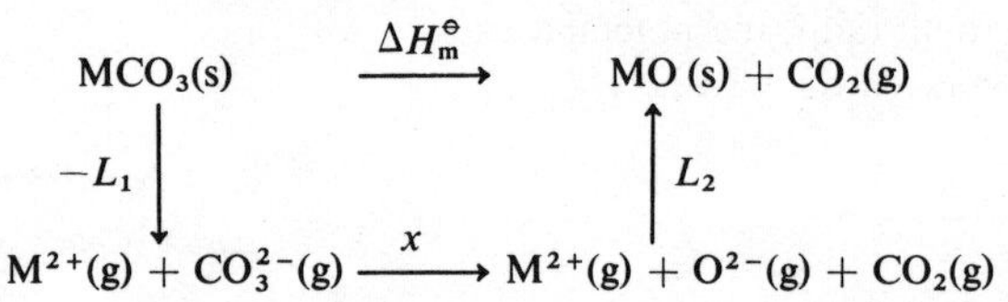

$$
\begin{array}{ccc}
MCO_3(s) & \xrightarrow{\;\Delta H_m^\ominus\;} & MO(s) + CO_2(g) \\[4pt]
{\scriptstyle -L_1}\Big\downarrow & & \Big\uparrow{\scriptstyle L_2} \\[4pt]
M^{2+}(g) + CO_3^{2-}(g) & \xrightarrow{\;x\;} & M^{2+}(g) + O^{2-}(g) + CO_2(g)
\end{array}
$$

Figure 5 Thermodynamic cycle for the decomposition of alkaline earth carbonates.

We first of all separate the M^{2+} and CO_3^{2-} ions to an infinite distance from one another. The standard enthalpy change here is very nearly $-L_1$ where L_1 is the lattice energy of the carbonate (Unit 5, Section 5.8.). Then we break the CO_3^{2-} ion up into $CO_2(g)$ and $O^{2-}(g)$:

$$CO_3^{2-}(g) \longrightarrow CO_2(g) + O^{2-}(g)$$

There will obviously be some enthalpy change here, and we call it x.

Finally, we bring the O^{2-} ions and M^{2+} ions together from an infinite distance apart to form a solid oxide MO. The standard enthalpy change here is very nearly equal to the lattice energy of the oxide, L_2.

According to the first law of thermodynamics, the total standard enthalpy change when reaction 3 occurs is the same whatever the reaction path. Thus, from the cycle,

$$\Delta H_m^\ominus = -L_1 + L_2 + x \qquad (4)$$

All the carbonates and oxides contain similarly charged ions and have the same formula type, so the values of C in equation 2 are the same for all the compounds. Substituting equation 2 into equation 4

$$\Delta H_m^\ominus = \frac{C}{r(M^{2+}) + r(CO_3^{2-})} - \frac{C}{r(M^{2+}) + r(O^{2-})} + x \qquad (5)$$

Now, it is not too easy to define exactly what is meant by the radius of a complex ion like CO_3^{2-}, but we obviously expect it to occupy more space in a structure than O^{2-} does. Thus presumably $r(CO_3^{2-})$ is considerably greater than $r(O^{2-})$.

Suppose that we are interested in the changes that take place in $\Delta H_m^\ominus$ as the cation M^{2+} changes from Be^{2+} to Ba^{2+}, that is, as $r(M^{2+})$ increases. In Figure 5, x is the same whatever the cation is, so in equation 5 the only terms which change when $r(M^{2+})$ is increased by a fixed amount are the two expressions derived from lattice energy terms:

12

$$\frac{C}{r(M^{2+}) + r(CO_3^{2-})} \quad \text{and} \quad \frac{C}{r(M^{2+}) + r(O^{2-})}$$

Both of these two expressions will drop.

Which drops by the greater amount?

The second. As $r(CO_3^{2-}) > r(O^{2-})$, a fixed increase in $r(M^{2+})$ produces a greater change in

$$\frac{C}{r(M^{2+}) + r(O^{2-})} \quad \text{than in} \quad \frac{C}{r(M^{2+}) + r(CO_3^{2-})}$$

If you find this hard to understand, try a numerical example putting, say, $C = 1$, $r(CO_3^{2-}) = 6$, $r(O^{2-}) = 2$ and changing $r(M^{2+})$ from 2 to 4.

Now both expressions drop, but in equation 5 the carbonate expression has a positive sign in front of it, and the oxide expression has a negative sign. Thus the changes tend to cancel one another because

$$\frac{C}{r(M^{2+}) + r(CO_3^{2-})}$$

goes down and

$$-\frac{C}{r(M^{2+}) + r(O^{2-})}$$

goes up.

However, we have seen that the increase in the latter is predominant so $\Delta H_m^{\ominus}$ becomes more positive as $r(M^{2+})$ increases. If we use the standard enthalpy change as a measure of stability, this means that the carbonates become more stable with respect to the oxide and CO_2 as the cation size increases. If you refer back to Unit 6, Section 6.3.1, you can confirm that this agrees with the experimental observations.

You can see then that a model which treats ions as charged spheres has possibilities in explaining the energy changes in certain types of chemical reaction. A rather sensational application of the model is described in the TV programme. However, our treatment left important questions unanswered. For example, the precise way in which lattice energies are affected by coordination numbers and lattice types we leave to a third level Course.

> SAQ 7 (Objective 6) In Unit 6, SAQ 27, we asked you which alkali metal fluoride you would choose if you wanted to prepare a compound containing the ClF_4^- ion by a reaction of the type
>
> $$MF(s) + ClF_3 = MClF_4(s)$$
>
> In the commentary we said that to judge from trends in the stabilities of other alkali metal compounds of this type, the stability of the compounds $MClF_4$ should increase down the group, and that therefore caesium fluoride would be a good choice.
>
> Explain the greater stability of $CsClF_4$ by using an ionic model.

7.3 Structures of the type MX_2

Until now, we have discussed structures of compounds with the formula MX. In all these structures, each particle is surrounded by particles of another kind and strong bonding forces run in all directions through the crystal. As we shall see, when we turn to compounds with the formula type MX_2, we not only encounter further examples of this kind of structure, but we also meet structures with new characteristics. The latter are especially interesting because, as you will discover, we see in them deviations from the properties expected of 'ionic' compounds. However, the first two structures that we shall consider resemble those of sodium chloride and caesium chloride in that each particle is surrounded by particles of the other type.

7.3.1 The fluorite structure

Calcium fluoride or fluorite provides an important example of a structure of the type MX_2.

Read Evans, Section 8.21, then try to answer the following questions.

> The coordination number of calcium is different from that of fluorine. Why is this?

As the structure is of the type MX_2, there must be twice as many fluorines as there are calciums. Thus if a calcium is surrounded by eight fluorines, a fluorine must be surrounded by only four calciums.

> SAQ 8 (Objectives 2 and 3) Draw a coordinate plan of the unit cell.

> SAQ 9 (Objectives 2 and 3) How many molecules of CaF_2 are there in the unit cell?

7.3.2 The rutile structure

The rutile structure is adopted by magnesium fluoride, and is pictured in Evans, Figure 8.05. All the unit cells that we have looked at so far have been cubic, but this one is tetragonal. This means that if we look along the x and y axes in Evans, Figure 8.05, we see rectangles, but if we look along the z axis we see a square. A plan of the structure looking down the z axis is shown in Figure 6.

In Figure 6 it is not easy to be as precise as we would like, because the x and y coordinates of the anions are not simple fractions as in the other structures that we have met. Indeed, they vary slightly from compound to compound. What you should remember is that the coordination is 6:3, made up of a nearly regular octahedron of anions around each cation, and a nearly equilateral triangle of cations around each anion.

7.3.3 The cadmium chloride and cadmium iodide structures

Until now, in all the structures that we have discussed, each particle detected by the X-ray method has been surrounded by particles of another kind, and strong bonding forces run in all directions through the crystal.

The cadmium chloride and cadmium iodide structures are not of this type. They are made up of layers which are themselves composed of a layer of cadmiums sandwiched between two layers of halogens. Part of one layer is pictured in Figure 7 in projection. The open circles with a thick surround represent atoms lying above the plane of the paper, and the open circles with a thin surround represent atoms lying below the plane. The red circles are in the plane of the paper. A three-dimensional drawing is shown on the cover of this book.

In these layers, the cadmiums are surrounded octahedrally by six halogens, but the halogens have three cadmiums on one side and *nothing on the other until the next layer is reached.* This is apparent from Figure 8 where parts of *two* adjacent layers are shown.

The difference between the cadmium chloride and cadmium iodide structures lies only in the relative disposition of the layers. We do not ask you to remember this difference, although if you are interested, it is described in Evans, Section 8.25. What you should remember is that, in both the $CdCl_2$ and CdI_2 structures, the nearest neighbours of any halogen in one direction are three other halogens in the next layer at a distance much greater than that which separates the halogen from its three cadmiums. This is clear from this Unit's Figure 8 where portions of two layers of the CdI_2 structures are shown. Notice that, as stated earlier in the Section, each layer of the structure is itself composed of a layer of cadmiums sandwiched between two layers of halogens.

Now this is no longer the arrangement that we would expect of an 'ionic' solid. If we described the structure in terms of ions, we should be forced to admit that instead of surrounding themselves as far as possible with cations, the halide anions have three cations on one side *and three anions on the other.* It seems,

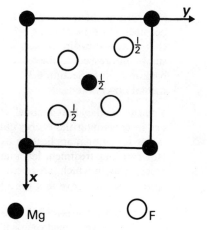

Figure 6 *Coordinate plan of the rutile structure of MgF_2.*

14

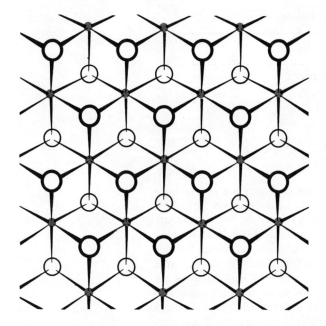

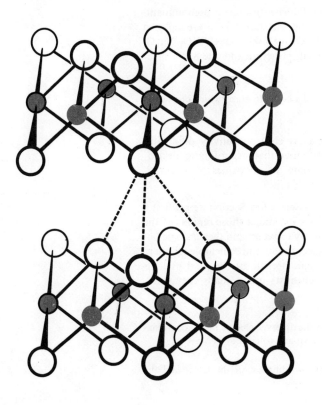

Figure 7 The structure of the layers in CdI$_2$ and CdCl$_2$. (Filled circles represent metal atoms in the plane of the paper; open circles, halogen atoms which lie in planes above and below that of the metal atoms.)

Figure 8 The halogen environment in the CdI$_2$ structure.

therefore, that in the structures discussed in this Section, we see a movement away from those characteristic properties that we associate with an ionic solid. Let us now see how some alkali metal and alkaline earth compounds distribute themselves among the structures discussed in Sections 7.3.1, 7.3.2 and 7.3.3.

7.3.4 The alkaline earth metal halides and alkali metal oxides

The structures of the halides of the alkaline earth metals are shown in Table 4. Those described as 'non-layer' are types that we have not described. All that you need to know is that they resemble the fluorite structure in that each particle is surrounded by particles of another type, and that the coordination number of the cation is even greater than the value of eight found in CaF$_2$. Those described as 'layer' have either the CdCl$_2$ or CdI$_2$ structures.

Table 4 Structures of alkaline earth metal halides

	F	Cl	Br	I
Mg	Rutile	Layer	Layer	Layer
Ca	Fluorite	Rutile	Rutile	Layer
Sr	Fluorite	Fluorite	Non-layer	Non-layer
Ba	Fluorite	Non-layer	Non-layer	Non-layer

Layer = CdI_2 or $CdCl_2$ structure.

The red dividing line shows clearly that the layer structures are concentrated towards the top right-hand corner of Table 4. For example, as we move from magnesium fluoride to magnesium iodide, we go from an 'ionic' structure to a layer structure between the fluoride and the chloride. Likewise as we move from barium iodide up to magnesium iodide, this transition occurs between strontium and calcium. Now you know from S100, Unit 8, that electronegativities decrease from fluorine to iodine and from magnesium to barium. Thus the concentration of layer structures in the top right-hand corner is in accord with the generalization made in S100, Unit 8: compounds with ionic characteristics are most likely when the electronegativity difference between metal and non-metal is large as in the bottom left-hand corner of Table 4.

Now, for the alkali metal oxides, M_2O, the *anti*-fluorite structure is predominant. This structure is exactly the same as that of fluorite, but the calciums are replaced by oxygens and the fluorides by alkali metals. Of the five oxides, Li_2O, Na_2O, K_2O, Rb_2O and Cs_2O, four have the anti-fluorite structure and one has a layer structure.

> If you were told that either Li_2O or Cs_2O had a layer structure, which one would you pick?

We expect you to choose Li_2O with the smallest electronegativity difference. However, the answer is wrong; Cs_2O, the compound that, according to electronegativity differences, should be the most 'ionic' of all oxides has an anti-cadmium chloride structure.

Thus in the microcosm of the compounds discussed in this Section we get first a confirmation of the ability of electronegativities to predict those regions of the Periodic Table where 'ionic compounds' are likely to occur, and then a single but crucial failure. This combination of overall success with unpredictable failure in important individual instances is a typical characteristic of the kind of half-theoretical and half-empirical concepts that chemists use. Usually they work, but occasionally they do not.

> SAQ 10 (Objective 5)   Can you account for the observed change in the coordination number of the cation as one moves down any column of Table 4?

> SAQ 11 (Objectives 2 and 7)   Describe the coordination of the metal and the oxygen in the structures of Cs_2O and Li_2O.

7.4   Structures of sodium, magnesium and aluminium chlorides

Implicit in the discussion of the structures of the alkaline earth metal halides was a study of the change in the structure of a fixed kind of halide as we moved down the alkaline earth metal group. We now examine changes of this kind as we move across the third period from sodium to chlorine.

We have looked carefully at the sodium chloride structure and seen that magnesium chloride has a layer structure of the $CdCl_2$ type. $AlCl_3$ also has a layer structure in the solid state.

> Within each layer the coordination of the aluminium is octahedral. What is the coordination number of chlorine?

As the stoichiometry of the chloride is 1:3, the coordination number of chlorine must be $6 \times \frac{1}{3} = 2$. As in the $CdCl_2$ and CdI_2 structures, the layers are composed of three tiers and each chlorine's neighbours lie all on one side (see Fig. 9). The open circles with a thick surround represent atoms lying above the plane of the paper and the open circles with a thin surround represent atoms lying below the plane. The red circles are in the plane of the paper.

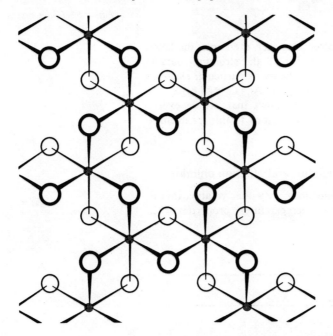

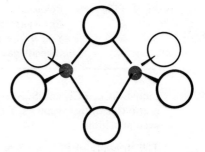

Figure 9 Structure of a layer of the $AlCl_3$ structure. (Filled red circles represent metal atoms in the plane of the paper; open circles, halogen atoms which lie in planes above and below that of the metal atoms.)

Now you know from Unit 3, Appendix 1, that molten $AlCl_3$ is not a conductor. This suggests that the melt does not contain ions. Moreover, in Unit 6, Section 6.2.9, we pointed out that at 300°C, the molecular formula of the vapour which you saw sublime in the TV programme for Unit 3 was Al_2Cl_6. All these observations have been clarified by electron diffraction measurements which show that in the vapour at temperatures below 400 °C, and in the melt, there are molecules of the type shown in Figure 10 where the coordination of the aluminium is tetrahedral. The structure and volatility of aluminium trichloride, and the non-conducting qualities of the melt are not the properties that we expect of an ionic compound.

In Figure 10, there are two types of bond: the terminal bonds formed by the four chlorines at the end of the molecule and the four bridging bonds formed by the two chlorines at the middle. These two types have different lengths.

Figure 10 The structure of Al_2Cl_6 molecules.

> Can you write a structure for the Al_2Cl_6 molecule in terms of Lewis-type two-electron bonds?

Possibly, if you have been unusually generous with the electrons. The first problem is that aluminium has only three electrons, and yet forms four bonds. In S100, and in this Course until now, we have only considered two-electron bonds in which each atom at either end provides one electron. Aluminium can form three of these, after which our electron count gives six electrons in the outer shell. This gaseous molecule, shown in Figure 11 (a), is formed when Al_2Cl_6 is heated to 600 °C.

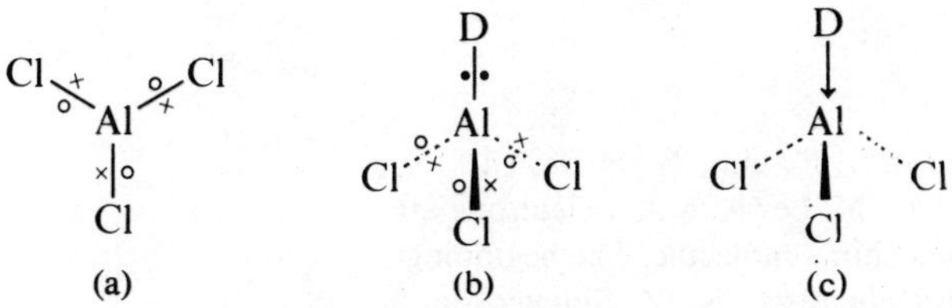

$$\text{(a)} \qquad \text{(b)} \qquad \text{(c)}$$

Figure 11 (a) Aluminium forms three 2-electron bonds. (b) Atom D with a lone pair completes the octet. (c) Representation of structure (b).

However, aluminium can attain an octet configuration and form four 2-electron bonds if another atom D is prepared to donate *a lone pair* as in Figure 11 (b). When this type of two-electron bond was first distinguished, it was called a dative bond, and given an arrow-like symbol which pointed from the donor to

17

the acceptor atom (Fig. 11 (c)). With this extension, we can salvage the Lewis theory and represent Al_2Cl_6 as shown in Figure 12.

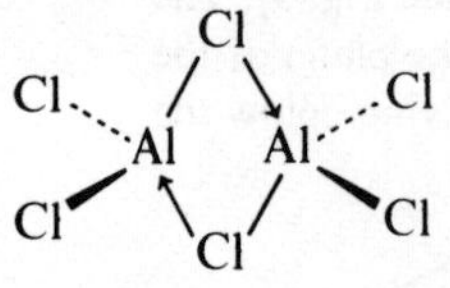

Figure 12 Dative bonds in Al_2Cl_6.

In this representation, every atom has an octet. However, even now, the Lewis theory requires further adaptation to account for all the structural data in Figure 10. The arguments expressed by the Lewis theory fix particular electrons in particular bonds, and imply that the two bonds in an $Al\text{—}Cl\text{→}Al$ bridge are different. However, the electron-diffraction study shows that, within experimental error, they have the same length. We shall return to this problem in Section 7.5.

7.4.1 Structures of the silicon, phosphorus, sulphur and chlorine chlorides

As the data in Table 5 show, all these compounds are rather volatile, that is, they have low boiling points or sublimation temperatures, properties often associated with covalent compounds.

Table 5. Properties of some chlorides

Compound	Description
$SiCl_4$	Colourless liquid boiling at 58 °C
PCl_3	Colourless liquid boiling at 76 °C
PCl_5	Colourless solid subliming at 163 °C
SCl_2	Red liquid boiling at 59 °C
SCl_4	Yellow solid at −80 °C, decomposing at −30 °C
Cl_2	Greenish gas at room temperature

Liquid $SiCl_4$, PCl_3 and SCl_2 together with Cl_2 contain discrete molecules; this is also true of their vapours. The nature of SCl_4 is uncertain because of its ready decomposition.

The structure of PCl_5 is interesting. An X-ray investigation of the solid shows that there are two types of grouping; one consists of a phosphorus surrounded tetrahedrally by four chlorines, and the other of a phosphorus surrounded octahedrally by six chlorines. The two groupings occupy the positions of the caesiums and chlorines in the caesium chloride structure. Solid PCl_5 is therefore assumed to be an ionic compound and is formulated as $PCl_4^+PCl_6^-$. However, it is obvious that such a structure in no way implies that the bonding between *phosphorus* and *chlorine* is ionic. For example, phosphorus has five outer electrons, so a P^+ ion has four. A PCl_4^+ ion could be generated if the P^+ ion formed four electron-pair bonds with four chlorine atoms when every element would attain an octet configuration. Moreover, in the vapour phase, PCl_5 consists of discrete molecules.

Thus, beyond aluminium, the chlorides of the third Period have structures consistent with bonding models which describe the bonds formed with chlorine as covalent.

7.4.2 Valence shell repulsion theory

Valence shell repulsion theory was introduced in S100, Unit 10, Section 10.4.1. It has a clear line of descent from the ideas of G. N. Lewis in that electrons are grouped in pairs which occupy fixed positions within a molecule. The beginnings of the theory were proposed by two Oxford chemists, N. V. Sidgwick and H. M. Powell, in 1940, and it is sometimes called the Sidgwick-Powell principle. The basic assumption is that the shapes of molecules are determined by the repulsion between outer or *valence* electrons which, as far as possible, are grouped in pairs. Reliable predictions are obtained only for molecules or ions of the *typical* elements.

SAQ 12 (Objective 8) Revise valence shell repulsion theory (S100, Unit 10, Section 10.4.1) by predicting the shapes of the molecules $SiCl_4$, PCl_3, PCl_5, SCl_2 and SCl_4 in the gas phase. The answer and comment for this question are given in the next Section.

To remind you of the theory, let us follow through its application to two molecules, BrF_5 and XeO_3.

1 Count the number of outer or *valence* electrons on the central atom. For the typical elements this is equal to the Group number. It is seven for bromine and eight for xenon.

2 Assign these electrons to the bonds.

To do this, an assumption about the nature of the bonds must be made. Presumably, each fluorine forms a single 2-electron bond which requires one electron from the fluorine and one from the central bromine. Thus five of the seven bromine electrons are used in the bonds.

Again, if we assume that oxygen forms double bonds with xenon to complete its octet, these bonds each require two electrons from the xenon, so the three bonds use six xenon electrons.

3 Divide the valence electrons that are not used in bonding into lone pairs as far as possible.

In BrF_5 subtraction of the five bonding bromine electrons from the total of seven leaves one lone pair. Likewise, in XeO_3, one lone pair is left.

4 Count each lone pair and each bond as a *repulsion axis*. The theory assumes that the repulsion axes get as far apart as possible, forming the shapes shown in Figure 13.

Figure 13 Shapes generated by two, three, four, five and six repulsion axes.

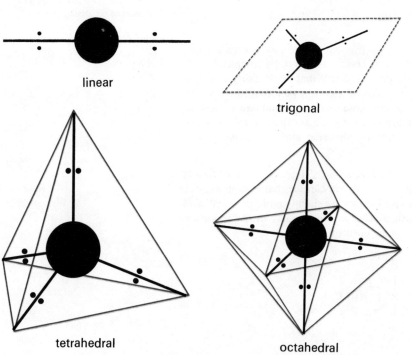

linear

trigonal

tetrahedral

octahedral

trigonal bipyramidal

For BrF_5, one lone pair and five bonds give six repulsion axes and Figure 13 tells us that these have an octahedral disposition. For XeO_3, one lone pair and three bonds give four repulsion axes and a tetrahedral disposition. Thus the predicted shapes are those in Figure 14, one of the axes being occupied by a lone pair in each case. This prediction is correct in that structural techniques show that the relative positions of the atoms are those given in Figure 14 (a) and (b). In 14(b) the middle oxygen is above, and the two others are below the plane of the paper.

You may also recall that minor deviations from regular shapes can often be accounted for by assuming that lone pair–lone pair repulsions exceed lone pair–bond pair repulsions which, in turn, exceed bond pair–bond pair repulsions. Thus in SCl_2 and PCl_3 which are based on four repulsion axes, the angles Cl—S—Cl

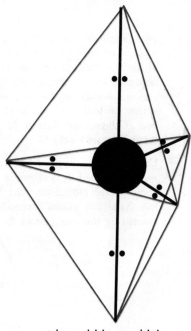

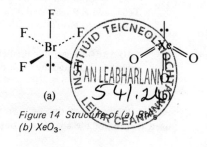

(a)

Figure 14 Structure of (a) BrF_5 (b) XeO_3.

and Cl—P—Cl are less than the value for a regular tetrahedron (109.5°) and we account for this by saying that the lone pair–bond pair repulsions are greater than the bond pair–bond pair ones. The shapes of $SiCl_4$, PCl_3 and SCl_2 are shown in Figure 15.

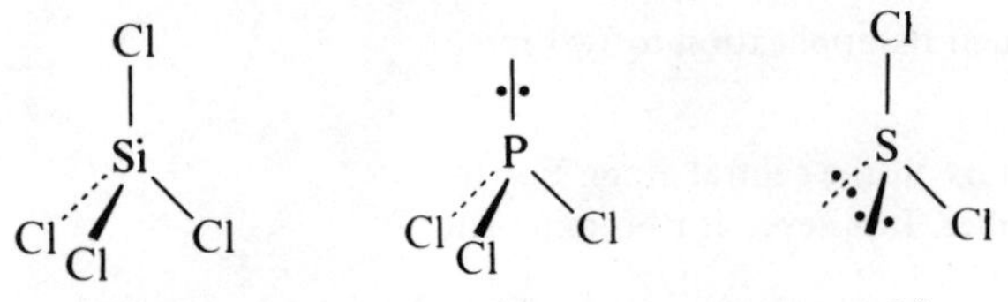

Figure 15 Structure of (a) SiCl₄ (b) PCl₃ (c) SCl₂.

Likewise, in BrF_5, the bromine atom lies slightly below the horizontal plane of the four fluorines (Fig. 14 (a)) because of the stronger repulsion between the lone pair and the four Br—F bonds.

We can now apply these rules to PCl_5 and SCl_4. They suggest that both molecules have five repulsion axes, one of the five being occupied by a lone pair in SCl_4. Figure 13 tells us that the five axes take up the shape of a trigonal bipyramid. We therefore predict that discrete PCl_5 molecules should have the shape shown in Figure 16 (a); this is correct. Now in the trigonal bipyramid, three of the positions called *trigonal* positions are different from the other two, called *axial* positions. These are marked differently on the PCl_5 structure in Figure 16.

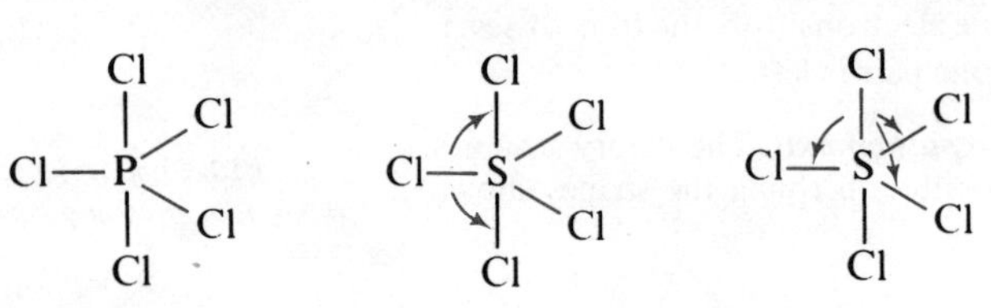

Figure 16 (a) Structure of PCl₅ with trigonal positions in red; (b) and (c) two possible structures for SCl₄ with dominant repulsions marked.

Now, the lone pair in SCl_4 could occupy a trigonal position as in Figure 16(b), or an axial position as in Figure 16(c). It is never possible to be completely sure of the preferred structure in cases like this, but a rule that works fairly well is to minimize the number of the strongest type of electron pair repulsions in the various alternatives. In this case, these are those lone pair–bond pair repulsions where the lone pair and bond pair are at 90° to one another. In (b) there are two of these, and three in (c), so (b) is the preferred alternative (the crucial repulsions are marked).

Unfortunately, we do not know the structure of SCl_4, but the shape of SF_4 is well established. It is shown in Figure 17, and you can see that it supports our choice of (b). Notice too, that the marked angles are slightly more than 180° and less than 120°, values consistent with a belief in the primacy of the repulsion exerted by the lone pair in the triangular position.

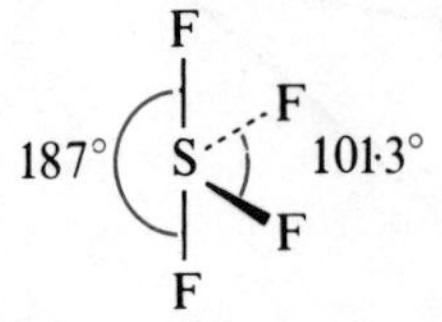

Figure 17 Structure of SF₄.

SAQ 13 (Objective 8) By heating the solids $BeCl_2$ and $SnCl_2$ to quite moderate temperatures, discrete $BeCl_2$ and $SnCl_2$ molecules can be obtained. What shape and bond angle would you expect these molecules to have?

SAQ 14 (Objective 8) By gently heating the solids $TeCl_4$ and XeF_4, discrete $TeCl_4$ and XeF_4 molecules can be obtained. What shape would you expect these molecules to have?

SAQ 15 (Objective 8) Predict the shape and bond angle of the sulphur dioxide molecule, SO_2, and the shape of the molecule $XeOF_4$ in which xenon is the central atom.

At the beginning of this Section we said that valence shell repulsion could be used to predict the shapes of ions of the typical elements as well as molecules.

In applying the theory to ions, only one slight modification is necessary: when you count the number of valence electrons in step (1), if the ion is an anion, add one electron for each negative charge; if the ion is a cation, subtract one electron for each positive charge. Having revised the number of valence electrons in this way, you proceed through steps 2–5 as before.

> SAQ 16 (Objective 8) Predict the shape of the ion ICl_4^- in the salt $KICl_4$. Recently, chemists prepared a compound $ClF_2^+SbF_6^-$. What shape and bond angle would you predict for the ion ClF_2^+?

Now that you have done SAQs 13–16, you should have a clear idea of the ability of valence shell repulsion theory to predict the shapes of molecules of the typical elements.

In the examples we have considered the theory is very successful. Nevertheless, if we look carefully, we can find cases where the theory fails. This seems to be particularly true at high temperatures. For example, at high temperatures, gaseous molecules of the alkali metal oxides, such as Li_2O, and of alkaline earth metal dihalides like $BaCl_2$ can be studied. As you can show for yourselves, valence shell repulsion theory predicts that Li—O—Li should be V-shaped, while Cl—Ba—Cl should be linear. In fact, the converse is true; Li_2O is linear and $BaCl_2$ is V-shaped.

Notice, too, that the theory only works in a limited area. It applies to molecules of the typical elements but not to those of the transition metals.

This reflects a general failing of classical theories: they have been built up inductively from a study of a limited range of compounds under everyday conditions and the resulting principles are inflexible. One advantage of this is that the principles lead to firm predictions, a quality not often shared by more advanced theories. However, the inflexibility sometimes becomes apparent when we examine compounds under abnormal conditions, for example, at high temperatures or when we try to extend the theories to a wider range of compounds. We sometimes find that then the classical principles no longer work.

7.4.3 Progression in chloride structures across the third Period

The structures described in Sections 7.4 and 7.4.1 show that when we examine the chlorides of the third Period from sodium to chlorine, we see a progression from an extended 'alternate-particle' chloride structure at sodium to discrete molecular chlorides at silicon, phosphorus, sulphur and chlorine. In between, at magnesium and aluminium chlorides, there are layer lattices with characteristics intermediate between those of extended alternate-particle and molecular structures. This progression also occurs in other Periods of the Periodic Table.

The types of bonding that we have considered until now have been ionic, covalent or metallic. In both this Course and S100, we have emphasized that the presence of ionic bonding in a compound is inferred from properties of the compound. In particular, an ionic compound conducts electricity in the molten state and has a structure in which particles of one kind surround themselves *in all directions* with particles or groups of particles of a different type. On the other hand, covalent compounds are distinguished by poor conductivity in the liquid state and often have structures in which discrete molecules can be picked out.

By such criteria, NaCl is an ionic compound, and the silicon, sulphur, phosphorus and chlorine chlorides are covalent. However, $MgCl_2$ conducts electricity in the molten state but has a layer rather than an alternate-particle structure, while $AlCl_3$, a non-conductor in the molten state, possesses a layer rather than a molecular structure. Thus we cannot classify $MgCl_2$ or $AlCl_3$ as ionic or covalent compounds; in fact, our rigid classification has broken down.

The natural response to this dilemma has been an attempt to attribute 'degrees of ionic or covalent character' to compounds and in the remaining part of this Section we shall consider two attempts to do this.

You are, in part, familiar with the first method. It uses electronegativity, the power of an atom to attract electrons to itself in a molecule, as a way of rationalizing the tendency of atoms to form ionic bonds (e.g. see S100, Unit 8, Section

8.4.9). G. N. Lewis was one of the first chemists to suggest this particular use of electronegativity although, as pointed out in Unit 5, Appendix 1 and in the radio programme referred to Section 7.1, to get some measure of electronegativity he had to use semi-qualitative observations on the tendency of atoms to form positive or negative states in chemical reactions. Since that time more sophisticated ways of assessing electronegativity have been devised and you will learn more about them in Unit 8.

They suggest that electronegativities increase across a row of typical elements. Consequently, for the chlorides of the third Period, the electronegativity *difference* between the atoms decreases as we move across the row from NaCl, and has fallen to zero by the time we come to Cl_2. Since the ionic character of the bond is likely to be greatest when the difference between the electron-attracting power of the atoms is greatest, this variation suggests that the chlorides should become less ionic and more covalent as we move across the row. As we have seen, this conclusion is in agreement with the observations that we have made on structural and physical properties. A similar successful use of electronegativity was pointed out in Section 7.3.4 when we discussed the structures of the dihalides of the alkaline earth metals.

In the same Section however, we noted the inability of electronegativities to cope with structural changes in the alkali metal oxides, and it turns out that this kind of failing becomes especially abundant when we compare compounds of different stoichiometric type. Thus silver chloride, AgCl, has a sodium chloride structure and conducts electricity in the molten state, although the electronegativity difference between the metal and the halogen is much less than in $AlCl_3$.

To combat this particular problem, a second method of rationalizing differences in ionic character was suggested in the 1920s by K. Fajans and the famous geochemist, V. M. Goldschmidt. It was concerned with the ways in which the cation in an ionic compound might affect the electrons around the anion through straightforward electrostatic forces. The process is pictured in Figure 18.

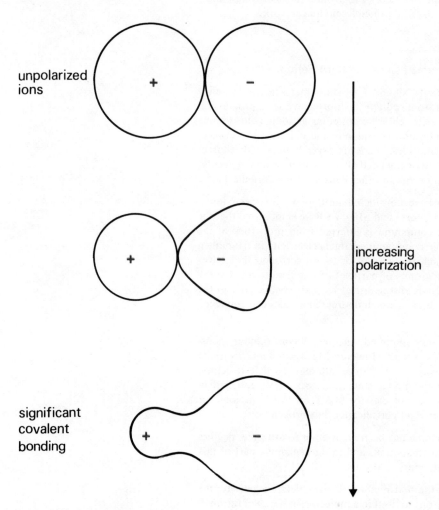

Figure 18 Ion polarization.

22

At the top of the figure, a positive and negative ion are juxtaposed as we might expect in our model of ionic compounds. However, as the cation becomes smaller, the electric field around it becomes more intense and it begins to pull the distribution of negatively charged electrons around the negative ion towards itself. At the bottom of Figure 18, the distortion has become severe, and the anion electron-distribution is shared between the two particles. Fajans and Goldschmidt associated this sharing with the presence of some covalent bonding.

The distortion of the anion pictured in Figure 18 is called polarization. The ability of a cation to cause polarization is known as 'polarizing power' and the willingness of the anion to be distorted is called its polarizability. In Figure 18, we associate high polarizing power with small cation size, but it should be obvious that a high cation charge should also be a cause of a large electric field and therefore of large polarizing power. Thus when we write down the formulae of $NaCl$, $MgCl_2$ and $AlCl_3$ in an ionic form, and consider the size and charge of the three cations, we find that the polarizing power of the cation increases from sodium to aluminium. This suggests that the covalent character of the chlorides should increase in the sequence $NaCl$, $MgCl_2$ and $AlCl_3$, and such a conclusion agrees with our experimental observations.

polarization
polarizing power
polarizability

> SAQ 17 (Objective 10) Can polarizing power and polarization account for the concentration of layer structures in the top right-hand corner of Table 4 on p. 16? What assumption do you have to make about the relation between the size and polarizability of the anion?

This SAQ and the preceding discussion show that the concept of ion polarization can be very useful. However, it too has flaws, and these often become apparent when we compare a typical element with an element from a different area of the Periodic Table. For example, our discussion has implied that cations of equal size and charge should polarize anions equally well.

Now, the ions Hg^{2+} and Ca^{2+} have similar sizes (see Table 3), but in $HgCl_2$, discrete molecules can be picked out, and the molten chloride is a very poor conductor of electricity. By contrast, $CaCl_2$ has the rutile structure and, when molten, conducts extremely well. This distinction between different areas of the Periodic Table (which is not predicted by our physical model of ion polarization) must be included in any full statement of the uses of polarization concepts (see rule 4, in Appendix 3 of Unit 10).

It is also interesting to note that ion polarization cannot account for the case of the alkali metal oxides discussed in Section 7.3.4.

These examples, and the realization that the concepts discussed in this Section lead to no *quantitative* conclusions, perhaps convey some idea of the difficulty that chemists have experienced in assigning degrees of ionic or covalent character to compounds. Ideally, one might hope that a solution of these difficulties would enable us to make statements like 'the bonds in $MgCl_2$ are 55.73 ± 0.02 per cent ionic' and then to use this figure to predict properties like structure, conductivity of the melt and volatility in a quantitative way. Nothing approaching this precision has ever been reached. Until it is, the words 'ionic' and 'covalent' are probably best regarded as labels for classes of compounds with defined properties that you should by now be familiar with. Chemists say a compound is ionic or covalent depending on how closely they feel it conforms to these properties.

It seems, therefore, that the hope, that a single theory of universal application could be built by marrying the concepts of ionic and covalent bonding in a way that can be precisely described, has not been fulfilled. Some new theory is necessary, and you will be introduced to the beginnings of one in the next Unit. Like Lewis's theory of covalent bonding, it uses the idea of two-electron bonds, but in a more fundamental way, and it has been able to surmount problems that the Lewis theory found hard to explain. In the remainder of this Unit, we describe some of those problems.

7.5 Resonance structures

If oxygen gas is exposed to ultra-violet light, a certain amount of the gas ozone is produced. This has the formula O_3 and its structure has been determined by microwave spectroscopy. It was found to be V-shaped with a bond angle of 117°.

Can you write a Lewis structure for the molecule?

An octet can be assigned to each atom with the structure in Figure 19 (a).

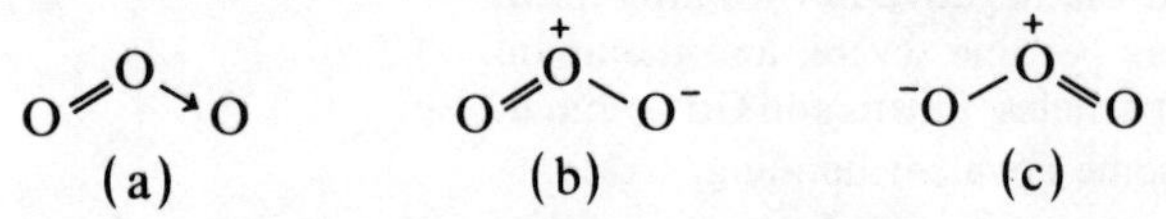

(a) (b) (c)

Figure 19 Lewis structures for ozone.

Figure 19 (a) also leaves a lone pair on the central oxygen to account for the bent shape on valence shell repulsion theory.

At this stage we introduce a more popular symbol for the dative bond. As pointed out in Section 7.13, the arrow symbol expresses the fact that *both* the electrons in the bond come from the central atom and count towards the valence electrons of the terminal oxygen. If we regard the molecule as composed of bound atoms, this picture implies that negative charge has been transferred from the central to the terminal oxygen. The new symbol puts a positive charge on the donor atom, a negative charge on the acceptor atom, and connects the two by a single bond (Fig. 19 (b)). It does not imply that in the real molecule, the central and terminal oxygens carry *integral* positive and negative charges—the representation should be regarded merely as a convenient symbol for the dative bond in which the electron pair is still shared by the two atoms.

What does Figure 19(a) or (b) imply about bond lengths in the ozone molecule?

As the bond types are different, the bond lengths should differ too. The very accurate microwave method, however, gave the figure 127.8 ± 0.2 pm for the lengths of both bonds. Now, this figure is different from the oxygen–oxygen bond length of 149 pm in gaseous hydrogen peroxide, HO—OH, and from the value of 121.1 pm in the oxygen molecule $O{=}O$. The figures for H_2O_2 and O_2 suggest that as the bond order increases from single to double, the bond length decreases. Many other examples could be quoted to show you that this seems to be a general rule: as the order of the bond between two particular elements increases, the bond length decreases. Consequently, the intermediate value in ozone suggests that the bond order is the same in both oxygen–oxygen bonds, and lies between one and two.

It may already have occurred to you that this failure of the Lewis theory may be related to the fact that we have only considered structure 19 (b) when 19 (c) seems equally likely. In fact, the observed geometry of the molecule seems like some kind of 'average' of the two.

This observation is related to an important weakness of the Lewis theory: in Lewis structures, electrons are static and, if they bind atoms together, are confined to particular bonds. But there is evidence that the electrons in atoms are in ceaseless motion. Thus the bonding electrons are not confined to particular bonds; they can range over the entire molecule. Consequently, the electron pair which gives the left-hand bond in Figure 19 (b) and the right-hand bond in Figure 19 (c) the higher order, contributes equally to both bonds. To retain Lewis structures and express this fact the following procedure was adopted. The possible Lewis structures are drawn out and connected by a double-headed arrow as in Figure 20.

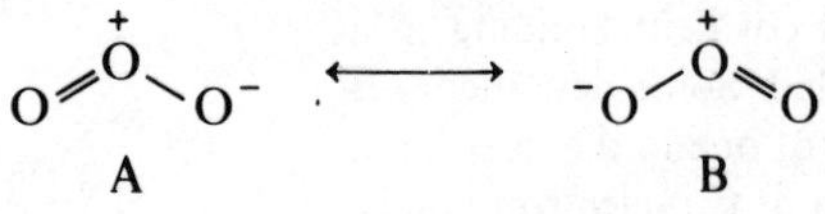

A B

Figure 20 Representation of ozone structure in terms of resonance hybrids.

Figure 20 in its entirety represents the ozone structure. Structures A and B never have a separate existence. Indeed Figure 20 was expressly designed to imply this by delocalizing those electrons which are confined to the double bond in either A or B over the whole molecule. Note that as the observed symmetrical molecule cannot be represented by A or B, its electronic state must be more stable (i.e. stable with respect to A or B).

Ozone is said to be a *resonance hybrid* of the orthodox Lewis structures A and B. Representations like those in Figure 20 enabled the Lewis theory to surmount the problems presented by ozone and similar structures. However, the method of approaching the real structure of a molecule via a number of structures which do not exist seems contrived, and the use of resonance hybrids has fallen into some disfavour in recent years. Nevertheless, as a way of accounting for the structures of some important molecules or ions, the resonance method is often easier to apply than other ways.

SAQ 18 (Objective 11)   Adjust the Lewis theory to the following observations using structures in which the outer shell of each atom contains eight electrons.

(a)   The C–C bond lengths in C_2H_6 and C_2H_4 are 154 pm and 134 pm, respectively. The carbon atoms in benzene, C_6H_6, make up a regular hexagon of side 140 pm.

(b)   The nitrogen-oxygen bond lengths in $\begin{array}{c}F\\ \diagdown\\ N=O\end{array}$ and $\begin{array}{c}HH\\ \diagdown\diagup\\ N-O\\ \diagup\\ H\end{array}$

are 113 pm and 147 pm, while the carbon-oxygen bond lengths in

$\begin{array}{c}Cl\\ \diagdown\\ C=O\\ \diagup\\ Cl\end{array}$ and CH_3OH are 117 pm and 143 pm.

The ions NO_3^- and CO_3^{2-} are Y-shaped with the N and C atoms symmetrically at the centre, and the N–O and C–O bond lengths are 122 pm and 129 pm.

(Represent negative charges on the ions by putting them on oxygens to form O^-; this can then make up its octet by forming single bonds.)

7.6   Stable electron configurations and electron pair bonds

In S100, bonding theories were firmly centred on the stability of the inert gas configuration. However, in the last two Units, we have continually encountered molecules or ions in which, according to the electron assignment of elementary bonding theories, atoms have taken up configurations different from those of the inert gases.

Name some of these molecules or ions.

Among these are AlCl (g) and $AlCl_3$ (g) (which are formed at high temperatures), PF_5, ClF_3, SF_6, BrF_5 and ICl_4^-. If two electrons are assigned to each bond, the central atom has four, six, ten or twelve outer electrons in at least one of these molecules.

Moreover, taken as a whole, the *transition* elements do not even show a very strong preference for any particular electronic configuration. For example, if the metals manganese, iron, cobalt, nickel and zinc are dissolved in dilute hydrochloric acid, dipositive ions are formed. These ions have the configurations [Ar] $3d^5$, [Ar] $3d^6$, [Ar] $3d^7$, [Ar] $3d^8$ and [Ar] $3d^{10}$ where [Ar] represents the electronic configuration of argon $1s^2 2s^2 2p^6 3s^2 3p^6$.

Obviously, the value of the concept of stable configurations diminishes sharply as the number of such configurations proliferates. G. N. Lewis clearly recognized this and, as we stated earlier, placed much more emphasis on the pairing of electrons to form chemical bonds. He remarked:

The rule of eight, in spite of its great importance, is less fundamental than the rule of two which calls attention to the tendency of electrons to form pairs. The electron pair, especially when it is held conjointly by two atoms, and thus constitutes the chemical bond, is the essential element in chemical structure.

As we saw in Section 6.1.8, the truth of this observation is forcefully demonstrated in the chemistry of the *typical elements*. Even here, however, we can find exceptions. At room temperature, nitric oxide is a colourless gas composed of single molecules with the formula NO. Each molecule contains 15 electrons. If we allow the oxygen atom two bonds, we find the nitrogen one electron short of an octet (Fig. 21 (a)).

From our experience of the typical elements, we would expect the odd ringed electron in Figure 21 (a) to represent unrealized bonding capacity. This could easily be remedied by pairing up the electron with that of another NO molecule to form the molecule (shown in Fig. 21 (b)) in which each atom exercises its customary valency. However, such pairing does not occur at normal temperatures and pressures.

Another odd electron molecule is the explosive gas, ClO_2. Valence shell repulsion theory assigns 7 outer electrons to the chlorine, 4 of which are required for the bonds to oxygen. This leaves one lone pair plus an odd electron.

If the latter is counted as a repulsion axis, an $O=Cl=O$ angle close to the tetrahedral value would be expected. The observed value is $116°$, but we can rationalize this high figure by assuming that the odd electron does not exercise as strong a repulsive influence as would a lone pair. Nevertheless, we cannot account for the unwillingness of the molecules to dimerize. This would pair up the odd electrons and allow the chlorine to be pentavalent as in the molecule ClF_5.

Thus even in the chemistry of the typical elements, we find that in some compounds, electrons are rather unwilling to be associated in pairs. Consequently, this re-emphasizes the need to go beyond the acceptance of the two-electron bond as a basic natural phenomenon.

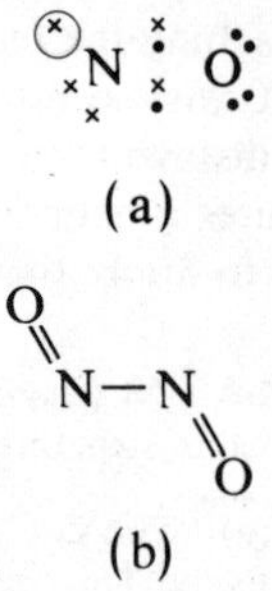

Figure 21 (a) Nitrogen with seven valence electrons in NO. (b) Molecule unknown at normal temperatures and pressures — each atom has an octet.

7.7 Summary and conclusions

In the early part of this Unit, we examined the structures of some compounds usually described as ionic, and showed how, by assigning ionic radii on the basis of a hard-sphere model, we can create a qualitative explanation of the variation of coordination number as the sizes of cations and anions change. You were also introduced to the beginnings of possible energetic applications of the ionic model.

We then showed that there were compounds such as magnesium and aluminium chloride whose characteristics straddled that collection of properties by which we recognize an ionic compound like NaCl, and that collection of properties by which we recognize covalent compounds like the chlorides of silicon, phosphorus, sulphur and chlorine. To account for this, it was necessary to find a concept that would describe and regulate the electronic changes that occur as one class of compounds changes into the other. We showed that electronegativity and the polarizing effects of ions could, in part, serve for this purpose, but we also emphasized the difficulties that have been experienced in assigning quantitative values to concepts like 'degree of ionic character'. In fact, these problems exposed the difficulties associated with the formulation of a comprehensive theory from the separate concepts of ionic and covalent bonding. Nevertheless, these concepts have achieved much, and we singled out for special attention Lewis' idea of electron pair bonds, because this has a clear relation to modern theories.

Although apparent exceptions to this pairing can be found in many transition metal compounds, and in one or two odd-electron molecules of the typical elements, it remains the central core of the elementary theory of covalent bonding, including valence shell repulsion theory, and survived adjustments such as the introduction of dative bonds and resonance hybrids. When a unified theory of chemical bonding finally emerged, its pedigree proved very different from that of previous bonding theories, although, as the great theoretical physicist, W. Heisenberg, remarked:

> . . . it seems questionable to me whether the quantum theory would have found or would have been able to derive the chemical results about valency if they had not been known before.

The achievements of the classical theory to which Heisenberg paid tribute culminated in G. N. Lewis's concept of the electron-pair bond which one modern theoretical chemist, J. W. Linnett, has called 'probably the most important and productive contribution that has ever been made to the subject of valency and chemical bonding'.

Appendix 1 (Black)

Radius ratio rules

In this Appendix, we try to apply the ideas in Section 7.1.6 in a quantitative way.

Now read Evans, p. 41. Remember that the figure is a picture of the vertical plane whose projection is the diagonal AC in our Figure 1.

Can you establish the formula at the bottom of p. 41 in Evans?

Turn to this Unit's Figure 1 (on p. 6). The limiting condition described by Evans in Figure 3.07(b), arises when the anions at the corners touch at the mid-point of the edges of the unit cell in Figure 1. If the cubic unit cell has side a, then the anion contact condition implies that $a = 2r_-$ where r_- is the radius of the anion. Now, by Pythagoras's theorem, $AC = \sqrt{2}a$, so the dotted rectangle in Evans, Figure 3.07(b) has sides a and $\sqrt{2}a$. Therefore, the rectangle diagonal is $\sqrt{3}a$ or, with the anion contact condition, $2\sqrt{3}r_-$. But the diagonal is also $2(r_+ + r_-)$ so,

$$2(r_+ + r_-) = 2\sqrt{3}\, r_-$$
$$r_+ = \sqrt{3}\, r_- - r_-$$
$$= r_-(\sqrt{3} - 1)$$
$$\frac{r_+}{r_-} = \sqrt{3} - 1 = 0.73$$

Once this condition is attained, according to our model, a decrease in cation size will no longer decrease the distance between the centres of the ions or, therefore, the attraction between the ions in the lattice. A structure with a lower coordination number, the NaCl structure, must therefore be adopted if r_+/r_- falls much below 0.73, and full cation–anion contact is to be maintained.

SAQ 19 (black)   Can you deduce a similar limiting condition for the NaCl structure?

Having completed SAQ 19, you should have deduced that if r_+/r_- for a halide MX falls much below 0.41, a structure of lower coordination number, presumably the zinc blende structure, should be adopted.

In Table 4 we give values of r_+/r_- for the alkali metal halides (r_-/r_+ where $r_- < r_+$ and contact is likely to occur between cations rather than anions).

Table 6 *Ion radius ratios for the alkali metal halides*

	Li	Na	K	Rb	Cs
F	0.51	0.75	1.00	0.91*	0.79*
Cl	0.37	0.55	0.73	0.81	0.92
Br	0.34	0.50	0.67	0.74	0.85
I	0.31	0.45	0.60	0.67	0.76

(r_+/r_- except for the starred cases where $r_- < r_+$)

In Table 6, those compounds with a radius ratio between 0.41 and 0.73 are enclosed by ruled lines. All of these have the sodium chloride structure so our calculations are partly successful. However, of the nine halides with a ratio greater than 0.73, only three have the caesium chloride structure, and of the three with a ratio less than 0.41, none has the zinc blende structure. Thus the sodium chloride structure is more prominent among the alkali metal halides than our radius ratio rules would suggest.

Evidently the coupling of a hard-sphere model with the assumption that cation–anion contacts are maximized is only partly successful.

SAQ 20 (black)   In Table 4, seven alkaline earth metal dihalides have either the rutile or fluorite structures in which the cation has a coordination number of six and eight respectively. Use Table 3 to find out how well the radius ratio rules work in this case.

SAQ answers and comments

SAQ 1   The answer is one. Suppose that the corners in Figure 1 are occupied by chlorines. Each chlorine at a corner contributes one-eighth of a chlorine to the unit cell and $8 \times \frac{1}{8} = 1$. This one chlorine is matched by one caesium at the centre.

SAQ 2

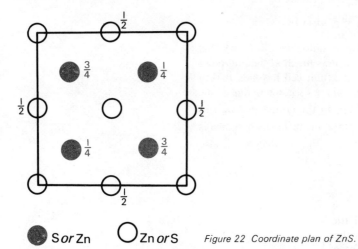

● $S\,or\,Zn$ ○ $Zn\,or\,S$ Figure 22 Coordinate plan of ZnS.

SAQ 3   The answer is four. If the zincs in Figure 22 are open circles, there are $6 \times \frac{1}{2} = 3$ at the centres of the faces and $8 \times \frac{1}{8} = 1$ at the corners. These four are matched by the four sulphurs entirely enclosed in the cell.

SAQ 4   From Evans, Table 3.01, the internuclear distance in LiI is 3.00 Å. Assuming that anion–anion contact occurs as in Figure 3, the iodide radius is $3.00/\sqrt{2}$ or 2.12 Å. This is 212 pm.

SAQ 5   From the internuclear distance in NaI, $r(\mathrm{Na}^+) = 3.23 - 2.12 = 1.11$ Å or 111 pm. Then from NaF, $r(\mathrm{F}^-) = 2.31 - 1.11 = 1.20$ Å or 120 pm. From the internuclear distances in RbI and RbF, $r(\mathrm{Rb}^+) = 1.54$ Å or 154 pm, so $r(\mathrm{F}^-) = 1.28$ Å or 128 pm.

SAQ 6   The value of C is constant for compounds which have the same formula type and which contain similarly charged ions. Thus C is the same for all compounds in group (ii) and for all compounds in group (iii).

In group (i) the compounds have the same formula type, MX, but the alkali metal halides contain singly charged ions and the alkaline earth metal oxides contain doubly charged ions. In group (iv), the compounds have different formula types and they contain differently charged ions.

SAQ 7   We are interested in changes in the value of $\Delta H_{\mathrm{m}}^{\ominus}$ for the reaction

$$\mathrm{MClF_4(s)} \rightarrow \mathrm{MF(s)} + \mathrm{ClF_3}$$

as the cation changes from Li^+ to Cs^+. The thermodynamic cycle to use is the following:

$$
\begin{array}{ccc}
\mathrm{MClF_4(s)} & \xrightarrow{\;\Delta H_{\mathrm{m}}^{\ominus}\;} & \mathrm{MF(s)} \quad + \mathrm{ClF_3} \\[2pt]
-L_1 \downarrow & & \uparrow L_2 \\[2pt]
\mathrm{M^+(g)} + \mathrm{ClF_4^-(g)} & \xrightarrow{\;\;x\;\;} & \mathrm{M^+(g)} + \mathrm{F^-(g)} + \mathrm{ClF_3}
\end{array}
$$

From this cycle

$$\Delta H_{\mathrm{m}}^{\ominus} = -L_1 + L_2 + x$$

where L_1 and L_2 are the lattice energies of solids $\mathrm{MClF_4}$ and MF respectively. Substituting the equivalents of equation 2:

$$\Delta H_{\mathrm{m}}^{\ominus} = \frac{C}{r(\mathrm{M}^+) + r(\mathrm{ClF_4^-})} - \frac{C}{r(\mathrm{M}^+) + r(\mathrm{F}^-)} + x$$

$r(\mathrm{ClF})_4^-$ exceeds $r(\mathrm{F}^-)$, so as $r(\mathrm{M}^+)$ increases, the decrease in the first term on the right-hand side is less than the increase in the term

$$-\frac{C}{r(\mathrm{M}^+) + r(\mathrm{F}^-)}$$

28

So because x remains constant, $\Delta H_m^\ominus$ increases as $r(\text{M}^+)$ increases. Thus the compound MClF_4 becomes more stable with respect to the fluoride and ClF_3 as the cation size increases, i.e. the stability order is $\text{CsClF}_4 > \text{RbClF}_4 > \text{KClF}_4 > \text{NaClF}_4 > \text{LiClF}_4$.

This means that the reverse preparative reaction

$$\text{MF} + \text{ClF}_3 = \text{MClF}_4$$

will be most favourable in a thermodynamic sense for the caesium compound, so this is the reaction to try.

SAQ 8

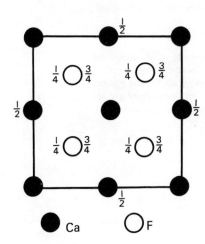

● Ca   ○ F

SAQ 9   The answer is four, made up of four calciums and eight fluorines.

SAQ 10   Down any column there is an increase in the coordination number of the cation. For example, the structure of the fluorides changes from the six-coordinate rutile structure to the eight-coordinate fluorite structure. Again, in the chlorides the coordination number is six at magnesium and calcium, eight at strontium and greater than eight in the non-layer structure of BaCl_2.

Table 3 (on p. 9) shows that the ionic radii of the cations increases down the alkaline earth metal group, so there is more room around each cation as one descends the group (see Section 7.1.6) and the coordination number can increase.

SAQ 11 Li_2O has the antifluorite structure. Thus the oxygen is surrounded by eight lithiums at the corners of a cube, while the lithium is tetrahedrally coordinated to four oxygens. In Cs_2O with an anti-CdCl_2 structure, each oxygen is octahedrally surrounded by six caesiums. Each caesium has three oxygens close to it on one side, and three caesiums in the next layer on the other side.

SAQ 12   The answer and comment for this question were given in Section 7.4.3.

SAQ 13   Beryllium has two valence electrons. Both are used up in forming the two bonds to chlorine. There are two repulsion axes so BeCl_2 should be linear. This is so. Tin has four valence electrons. Two are used up in forming the two single bonds to chlorine, and this leaves one lone pair. There are three repulsion axes which should therefore be disposed in a triangular sense (see Fig. 13), so SnCl_2 should be V-shaped with a bond angle sightly less than 120° because of the primacy of the lone pair–bond pair repulsions. Experimentally, this is found to be the case.

SAQ 14   Tellurium has six valence electrons like sulphur, so the shape of TeCl_4 can be estimated by the same reasoning used for SCl_4 in Section 7.4.2. The shape obtained is shown in Figure 17 (on p. 20). An electron-diffraction study confirms this, and also shows that as in SF_4, the angles marked in Figure 17 are slightly more than 180° and less than 120°.

Xenon has eight valence electrons, and four are used up in the Xe—F bonds, leaving two lone pairs. There are six repulsion axes with an octahedral disposition and two possible shapes:

The strongest repulsions are the lone pair–lone pair ones, and these can be minimized by putting the lone pair axes at 180° as in (b) rather than at 90° as in (a). Thus we predict a planar XeF_4 molecule as in (b). This is confirmed by experiment.

$$
\begin{array}{cc}
\text{F} & \\
\text{F}-\text{Xe}-\text{F} & \text{F}~\overset{\bullet\!\bullet}{}~\text{F} \\
 & \overset{\diagdown}{\text{Xe}}\diagup \\
\text{F}~\overset{\bullet\!\bullet}{} & \text{F}~\overset{\bullet\!\bullet}{}~\text{F} \\
\text{(a)} & \text{(b)}
\end{array}
$$

SAQ 15 Sulphur has six valence electrons, and four are used up in forming two double bonds to oxygen. This leaves one lone pair giving three repulsion axes. We predict a V-shaped molecule with a bond angle close to 120°. The experimental value from microwave spectroscopy is 119.5°. In XeOF$_4$, two of the xenon electrons are involved in the double bond to oxygen, and four in the Xe—F bonds leaving one lone pair. There are six repulsion axes, and as we assign four electrons to the Xe=O bond, we might expect the repulsions between this bond and the lone pair to be the greatest. The structure that

minimizes this repulsion is

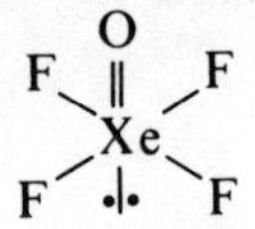

and this is confirmed by experiment.

SAQ 16 When applying valence shell repulsion theory to ions like ICl$_4^-$, count the charge towards the valence electrons of the central atom. Seven electrons on the iodine plus one from the negative charge gives eight valence electrons in all for the iodine. Four of these are used in the I—Cl bonds, leaving two lone pairs. Thus there are six repulsion axes distributed octahedrally, and there are two alternative shapes (a) and (b).

(a)   (b)

The lone pair–lone pair repulsions are the most significant and (a) is the structure that minimizes these; (a) is also the observed structure.

Chlorine has seven electrons, but we deduct one for the positive charge on ClF$_2^+$ leaving six. Two are used up in the bonds to fluorine and two lone pairs are left. There are four repulsion axes with a tetrahedral distribution, so we predict a V-shaped ion with a bond angle slightly less than the tetrahedral value because of the primacy of the lone pair–bond pair repulsions. Experimental evidence suggests that this is probably so in ClF$_2^+$ SbF$_6^-$.

SAQ 17 Yes they can. Down any column, the size of the cation increases so, as the charge is constant, we expect the polarizing power of the cation to decrease. This suggests the structures should become more ionic, and this agrees with the replacement of layer structures by those in which each particle is surrounded by those of another type as we descend the chloride, bromide and iodide columns.

Across any row, the cation is fixed, but the size of the anion increases. If we argue that the larger the anion, the less strong is the attraction of its nucleus for its outer electrons, then we can also add that those outer electrons should be more easily distorted by a cation. Consequently, the polarizability of the anions should increase across the row, and covalent structures should become more prominent. Such a transition is apparent in the magnesium and calcium rows where there are layer structures to the right.

The two effects we have discussed, together account for the concentration of layer structures in the top right-hand corner of Table 4 (on p. 16). Our assumption, of course, is that the polarizability of anions increases with size. Physical measurements confirm this (see Unit 10).

SAQ 18 (a) The observed bond length in benzene suggests that the carbon–carbon bond order lies between that in ethane and that in ethylene, i.e. between one and two. A representation (right) which incorporates the equivalent resonance structures, A and B, accounts for this.

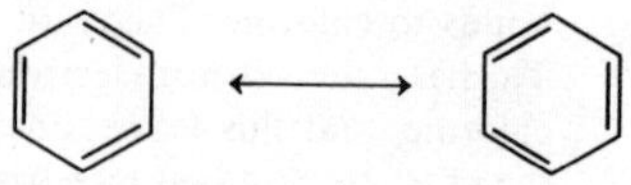

A   B

(b) Again, the bond lengths suggest a bond order of between one and two in the nitrate and carbonate anions. If we use singly bonded O$^-$ ions to carry the charges, basic Lewis structures in which all atoms have an octet are:

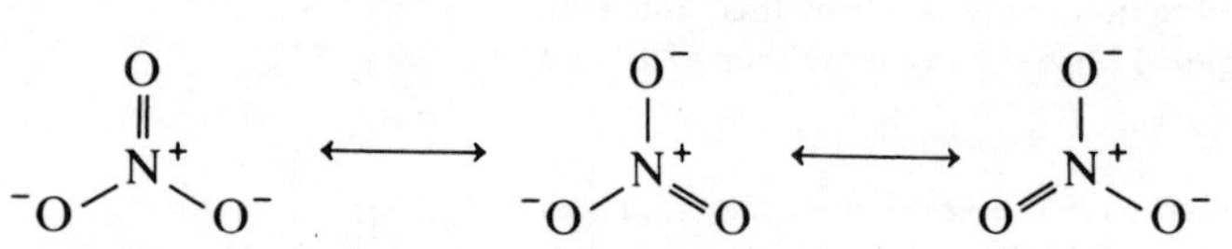

and

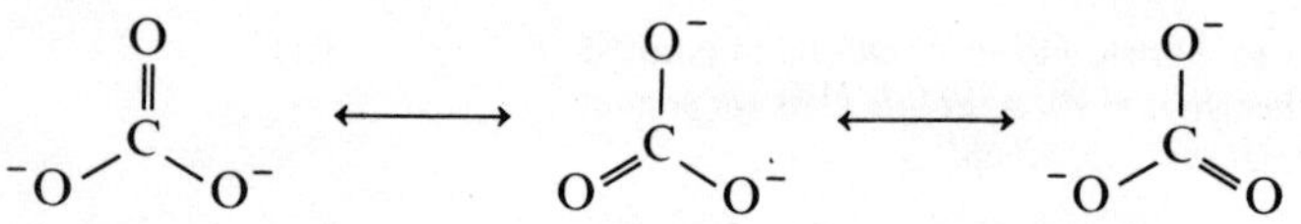

30

The fact that all bond lengths are equal can be represented by resonance hybrids based on these structures.

Here the N → O dative bond has been altered to the more modern symbol. The representation suggests that bond orders lie between one and two.

SAQ 19 (black) Look at Evans, Figure 3.02. The horizontal plane of the octahedron is now the one in which anion–anion contact will first occur if the cation size is steadily decreased. At the contact condition, the square in Figure 23 has side $2r_-$ and by Pythagoras's theorem a diagonal of $2\sqrt{2}r$.

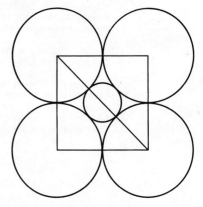

$$\text{Thus } 2\sqrt{2}r_- = 2(r_+ + r_-)$$
$$\sqrt{2}r_- - r_- = r_+$$
$$\frac{r_+}{r_-} = \sqrt{2} - 1$$
$$= 0.41$$

Figure 23 Anion–anion contact condition in the NaCl structure.

SAQ 20 (black) In MX_2 halide structures, the cation has the higher coordination number, and is usually smaller than the anion. Thus the limiting condition for anion–anion contact around the cation must be examined. In the fluorite structure, the cation coordination is the same as Cs in CsCl, while in the rutile structure it is the same as Na in NaCl. The hard-sphere model therefore suggests that the fluorite structure should occur when $r_+/r_- > 0.73$, and the rutile structure when $r_+/r_- > 0.41$. Of the seven compounds to be examined, only the structure of $SrCl_2$ disagrees with this prediction. Thus the model scores a higher success rate with the alkaline earth metal halides than with the alkali metal halides.

Acknowledgement

Figures 7, 8 and 10 are based on illustrations in A. F. Wells (1962) *Structural Inorganic Chemistry*, Oxford University Press.

S25- Structure, Bonding and the Periodic Law